Primitive sensory and communication systems

Primitive Sensory and Communication Systems

The Taxes and Tropisms of Micro-Organisms and Cells

Primitive Sensory and Communication Systems

The Taxes and Tropisms of Micro-Organisms and Cells

Edited by

M. J. CARLILE

Department of Biochemistry
Imperial College of Science and Technology
London

1975

ACADEMIC PRESS

London New York San Francisco

A Subsidiary of Harcourt Brace Jovanovich, Publishers

ACADEMIC PRESS INC. (LONDON) LTD.
24/28 Oval Road
London NW1

United States Edition published by
ACADEMIC PRESS INC.
111 Fifth Avenue
New York, New York 10003

Library of Congress Catalog Card Number: 75-19621
ISBN: 0-12-159950-7

Text set in 11/12 pt. Monotype Baskerville, printed by letterpress,
and bound in Great Britain at The Pitman Press, Bath

Contributors

Julius Alder *Departments of Biochemistry and Genetics, University of Wisconsin, Madison, Wisconsin, U.S.A.*

M. J. Carlile *Department of Biochemistry, Imperial College of Science and Technology, London SW7 2AZ, England*

Graham W. Gooday *Department of Biochemistry, University of Aberdeen, Aberdeen, Scotland*

Theo M. Konijn *Cell Biology and Morphogenesis Unit, Zoologisch Laboratorium der Rijksuniversiteit te Leiden, Leiden, Kaiserstraat 63, The Netherlands*

W. Nultsch *Department of Botany, University of Marburg, Germany*

P. C. Wilkinson *Department of Bacteriology and Immunology, University of Glasgow, Scotland*

Preface

Some years ago I decided to write a book on what seemed to me the promising but rather neglected topic of the taxes and tropisms of micro-organisms and cells. However, soon after I had commenced writing the neglect ceased and the promise began to be fulfilled. The volume of work on phototaxis and phototropism increased. The bacterium, *Escherichia coli*, so extensively utilized by geneticists and biochemists, was exploited in studies which began to reveal the molecular basis of chemotactic behaviour in bacteria. Acrasin, the attractant in morphogenesis in the cellular slime mould *Dictyostelium discoideum*, was shown to be cyclic AMP and progress was made in understanding its action. Several fungal and algal sex attractants were discovered, and there was a resumption of activity in the study of the chemotactic responses of mammalian leucocytes.

The ensuing rapid progress in these varied (although related) fields soon made it clear that an account that was both critical and up-to-date was unlikely to be achieved by a single author. I therefore sought the help of those involved in the most active areas of research, and received their enthusiastic cooperation in the production of appropriate chapters. My only regret is that I was unable to include a chapter on phototropism. However, the most intensive work on phototropism is being carried out on Phycomyces and the massive review by Bergmann *et al.* and the other papers on Phycomyces cited in Chapter 1 constitute an introduction to this field.

I thank my five co-authors for their cooperation—their chapters were a pleasure to edit and will constitute a valuable introduction to the rapid progress now being achieved. I thank also the staff of Academic Press for their efficiency and helpfulness.

M. J. Carlile
August 1975

Contents

Chapter 1

Taxes and tropisms: Diversity, biological significance and evolution

M. J. CARLILE

Department of Biochemistry,
Imperial College of Science and Technology, London, England

I. Introduction

Micro-organisms and cells show sensory responses to a wide range of environmental stimuli, including signals from other micro-organisms and cells. These responses take the form of oriented growth or of movement, and are conveniently divided into *tropisms* and *taxes*.

Tactic responses (*taxes*, sing. *taxis*) occur in most mobile cells or organisms in response to a variety of stimuli. A positive taxis is one in which movement towards the source of the stimulus occurs and a negative taxis is one in which there is movement away from the stimulus. The term taxis has been used by some workers in a very broad sense and by others in a more restricted way. The narrower usage was encouraged by a classic discussion of oriented movements (Fraenkel and Gunn, 1940, reprint with additional notes and supplementary bibliography, 1961). They classified the oriental movements of animals into kineses, taxes and transverse orientations, of which only the first two need concern us here. Kineses are displayed when organisms are incapable of detecting the direction of a gradient or of the source of a stimulus. Instead, they respond to a change in the intensity of the stimulus by a changed rate of locomotion (orthokinesis) or turning (klinokinesis) in such a way as to lead to net movement towards or away from the source of the stimulus, even though the organism is incapable of precisely oriented movements. Fraenkel and Gunn confine use of the term taxis to responses in which the organism becomes oriented with respect to the stimulus source and travels in a more-or-less direct way towards or away from the source of the stimulus. They further subdivided taxes on the basis of the mechanism of the response (see section III of this chapter). The classification of oriented responses by Fraenkel and Gunn had the merit of bringing some order and logic into existing information and stimulating thought on the mechanism of response. However, a system of classification of responses based on the mechanism of aggregation has a grave practical defect, namely that it is often clear that a response exists many years before there is any hard information about its mechanism. This is especially true with micro-organisms. For example, the "chemotaxis" of bacteria to nutrients was the subject of years of research and of many fine papers (see Ch. 3) long before there was any evidence as to whether the response was a taxis or a kinesis. Hence the Fraenkel and Gunn classification can only be applied to a small fraction of the organisms showing oriented responses since it demands generally non-available information. It is perhaps this that has led many workers to ignore the scheme and to continue to use, for example, chemotaxis to cover "chemo-aggregation" and "chemo-dispersal" regardless of mechanism. Moreover kinesis is used by some workers (e.g. Nultsch, Chapter 2) to describe an effect on the rate of locomotion (e.g. photokinesis) regardless of whether aggregation results. In this volume we shall use taxis in the broader sense except where otherwise indicated. Thc different types of oriented movement are summarized in Table 1.

TABLE I. Terms used in describing the effects of various stimuli on the movement and growth of micro-organisms.

Effect	Result	Common Terms
1. Rate of movement of motile cells is influenced	May perhaps lead (Fraenkel and Gunn, 1961) to the accumulation of organisms through a decreased, or their dispersal through an increased, rate of movement.	Orthokinesis (Fraenkel and Gunn) or simply kinesis.
2. Frequency of turning of motile cells is influenced	This, certainly if accompanied by adaptation, could lead to net movement away from the source of a stimulus through increase of turning when an increased stimulus is encountered, or towards it, if the frequency of turning is greater as a cell travels down a stimulus gradient.	Klinokinesis (Fraenkel and Gunn), or phobotaxis (Kühn[1]).
3. Orientation of motile cells is influenced.	Cells steer towards or away from the source of a stimulus.	Topotaxis (Kühn) or "true taxis". The only taxis recognised by Fraenkel and Gunn. Subdivisions are recognised by both Kühn and Fraenkel and Gunn.
4. Direction of growth of an immobile cell is influenced.	Growth towards or away from the source of a stimulus	Tropism.

[1] The classification adopted by Kühn is given in a Table in Fraenkel and Gunn (1961).

Tropisms (tropic responses) are common in organisms which are firmly attached to a substratum and hence are incapable of locomotion and taxis. Orientation of part of the organism in response to a stimulus occurs through oriented growth. As with taxis, tropism may be positive or negative. The term tropism was used in a very broad way by Jacques Loeb, in a classic book on animal orientation (Loeb, 1918, reprinted 1973), to include taxes and indeed all oriented responses in which there was no evidence of volition; this usage has not been widely accepted, certainly not by those concerned with micro-organisms and cells.

This chapter and indeed the whole book will be selective in restricting discussion mainly to responses which have been the subject of considerable investigation or are clearly of major importance. A massive review of the earlier literature of taxes and tropisms in plants and micro-organisms is provided by Ruhland (1959, 1962) and has sections on phototaxis and phototropism, chemotaxis and chemotropism, geotaxis and geotropism, thermotaxis and thermotropism, electrotropism, thigmotropism, and hydrotropism. An early work that remains of interest is the book by Jennings (1904, republished 1962), *Behaviour of the Lower Organisms*, many of the tactic responses he described being ripe for further investigation.

II. The nature of the stimulus

Micro-organisms and cells respond to a wide range of physical and chemical stimuli. The nature of the stimulus constitutes a convenient basis for classifying taxes and tropisms.

A. Light

Light is essential for the survival of the many micro-organisms that derive their energy from this source (photosynthetic bacteria and algae) and for other micro-organisms it is one of the most consistent and reliable indicators of direction. It is therefore not surprising that sensory responses to light are widespread and the subject of much research. In this volume phototaxis is discussed by Nultsch (Chapter 2), so further discussion here is unnecessary.

Positive phototropism is widespread in the spore bearing structures (sporangiophores and conidiophores) of fungi (Carlile, 1970a) and negative phototropism occurs with germinating spores of a wide range of organisms, including some plant pathogenic fungi (Carlile, 1966, 1970a). The most intensive studies on phototropism have been with the giant sporangiophores of the fungus *Phycomyces* and a massive review of work with this species has been published (Bergman *et al.*, 1969). More recent studies on the light sensitivity and phototropism of the *Phycomyces* sporangiophore include those of Dennison and Bozof (1973), Foster and Lipson (1973), Jan (1974), Jesaitis (1974), Johnson and

Gamov (1972), Magnus and Wolken (1974), Ortega and Gamov (1970) and Petzuch and Delbrück (1970).

B. Chemical agents

Whereas light is of great importance to many organisms, the chemical environment is of importance to all micro-organisms and cells. In this volume Adler discusses chemotaxis of bacteria (Chapter 3), Konijn the chemotaxis of amocboid cells (Chapter 4), Gooday the very diverse chemotactic and chemotropic responses of algae and fungi (Chapter 6) and Wilkinson the chemotaxis of leucocytes (Chapter 6). This by no means exhausts the diversity and significance of chemotactic and chemotropic responses, although it does cover the areas that have been most thoroughly studied in recent years. Some areas that have been inadequately studied but seem ripe for further investigation will be mentioned briefly.

The most elaborate form of cellular interaction and morphogenesis seen in prokaryotes occurs in Myxobacteria in which cells come together to form a fruiting body visible to the naked eye, a morphogenetic process mimicking that in the Acrasiales (Chapter 4). This response was the subject of a preliminary study by Bonner (1952) on *Chondromyces crocatus* and led him to suspect that both contact guidance and chemotaxis were involved. Subsequently McVittie and Zahler (1962) provided "fairly convincing evidence" for chemotaxis in the fruiting of *Myxococcus xanthus* and Fluegel (1963) for *Myxococcus fulvus*. Now that an increasing range of Myxobacteria are being cultured (Peterson, 1969) and are being used for studies on cell interaction and morphogenesis (Dworkin, 1973), further study of chemotaxis in this group is desirable.

There has been little work in recent years on the chemotactic responses of Protozoa other than amoeboid forms. Dryl (1973) has reviewed chemotaxis of ciliates, discussing mainly taxis to the optimum pH and away from high concentrations of salts and alcohols.

It is fair to say that the study of the chemotaxis of the motile gametes of the lower green plants, other than Algae, has suffered from a half-century or more of neglect. Early workers demonstrated chemotaxis of spermatozoa to the female organs in a range of liverworts, mosses, ferns and other Pteridophytes and in some instances showed taxis to various chemicals. These studies were reviewed by Machlis and Rawitscher-Kunkel (1963). Little has been done on the topic since Rothschild (1956) studied the chemotaxis of a range of Pteridophyte spermatozoa to various organic acids and Brokaw (1958a, b) showed that precise positive tactic orientation of Bracken (*Pteridium aquilinum*) spermatozoa occurred with a gradient of bimalate ions or a voltage gradient in the presence of bimalate ions, and required the presence of calcium ions (Brokaw, 1974)—an interesting finding published after a delay of 15 years.

The chemotropism of the pollen-tubes of higher plants has not been quite so badly neglected, yet relatively little work seems to have been done on this topic since the important studies of Machlis, Mascarenhas and Rosen about ten years ago. Early work on the pollen tube chemotropism and its probable role in guiding the pollen tube through the stigma and style towards the ovary was reviewed by Mascarenhas and Machlis (1962a) and Rosen (1962, 1968) and the story was brought up to date by Mascarenhas (1973). The pollen tubes of *Antirrhinum majus* were found to show positive chemotropism to calcium ions (Mascarenhas and Machlis, 1962b, 1964), as did *Narcissus pseudonarcissus* and *Clivia miniata*. However, the pollen of *Lilium longiflorum*, which shows strong chemotropism to the female reproductive structures of its species and to extracts from them, does not respond to calcium. The present evidence indicates that calcium is a chemotropic factor in fertilization in some species but is perhaps not the only one, and is without effect in others.

Chemotaxis of animal spermatozoa to female gametes undoubtedly occurs in a range of hydroid Coelenterates (Miller, 1966, 1973); active extracts were obtained. Whether sperm chemotaxis occurs in other animals remains controversial; it is not clear whether the absence of proof reflects an absence of the phenomenon or technical difficulties.

C. *Water*

Hydrotropism is probably of importance in fungi. Spores of many plant pathogens can germinate in the absence of liquid water and positive hydrotropism may guide germ-tubes of spores deposited on leaves through stomata into host tissues (Dickinson, 1960). The dispersal of fungal spores generally requires that the spore-bearing structures should grow away from a moist substratum into the relatively dry air. However, very little work has been done on the topic, probably due to technical difficulties in obtaining and measuring humidity gradients. It is thought (Johnson and Gamov, 1971) that the ability of the *Phycomyces* sporangiophore to avoid solid objects is probably a negative hydrotropism.

D. *Gravity*

Geotropism is widespread in fungi (Banbury, 1962). The most detailed recent studies have been on the negative geotropism of the sporangiophore of *Phycomyces*. It was found that gravity acted on the sporangiophore in two ways (Dennison, 1961). When gravity causes a passive bending of a sporangiophore, the lower side is compressed and the upper side stretched. The stretch receptors (Dennison and Roth, 1967) respond with increased growth on the lower side and reduced growth on the upper, and since this growth is intercalary, negative geotropism results. In addition, a negative geotropism is elicited through a second pathway, probably involving the upward displacement of the vacuole by gravity (Banbury and Hankinson, personal communication), which leaves more protoplasm adjacent to the lower than the upper side of the sporangiophore, and leads in some way to more growth on the lower side.

E. Contact and pressure

A range of taxes, tropisms and oriented responses are elicited in micro-organisms and cells by pressure, which may or may not involve contact, and contact, which may or may not involve pressure.

A response due to pressure following contact is the so-called avoidance reaction (in fact a collision reaction) in flagellates and ciliates. A true avoidance reaction, that of the *Phycomyces* sporangiophore, was mentioned above (section II C). The avoidance reaction has been studied in detail in *Paramecium*, which when it collides with a solid object reverses, and if prodded from behind speeds up. The collision leads to a stretching of the anterior membrane, which initiates a series of events leading to reversal of ciliary beat and backward swimming (Eckert, 1972).

In the *Phycomyces* sporangiophore stretching leads to a transient decrease and compression to a transient increase in growth rate (Dennison and Roth, 1967). If the sporangiophore is bent, then compression of one side and stretching of the other result in differential growth which will rectify the consequences of the deformation. As indicated above (section II D), the stretch receptors have a role in negative geotropism.

Response to pressure without contact with a solid object is seen in the positive rheotaxis (up-stream swimming) of human and bull spermatozoa (Bretherton and Rothschild, 1961) and of some fungal zoospores (Katsura and Miyata, 1971). Apparent *rheotropism* occurs in fungal germ tubes, upstream growth having been found in *Rhizopus nigricans* by Stadler (1952) and downstream growth in *Botrytis cinerea* by Muller and Jaffe (1965). Here, however, rheotropism is not due to pressure, but to high concentrations of a growth retarding (*R. nigricans*) or growth stimulating (*B. cinerea*) agent on the downstream side of the spores.

A widespread phenomenon that has been known by a variety of names is *contact guidance*. Early in the century a number of investigators (e.g. Kufferath, 1911) studied the oriented growth of the bacterium *Kurthia zopfii* on the surface of gelatin media. They showed that growth occurred parallel to lines of stress in the medium. The effect was sometimes termed *elasticotropism*. Stanier (1942) demonstrated *elasticotaxis* in the myxobacterium *Myxococcus fulvus* and related species, and later

(Stanier, 1947) in the Nostocaceae, a group of cyanobacteria (blue-green algae): cell migration again occurred parallel to lines of stress in agar media. According to Weiss (1945) contact guidance also determines the direction of growth of nerve fibres in stressed coagulated plasma, growth occurring parallel to the applied stress, probably guided by oriented fibrin molecules or micelles. However, Dunn (1973) has concluded that the parallel orientation of nerve fibres in tissue cultures is due largely to a different form of contact guidance—exploratory filopodial processes which are produced by the growing nerve fibre are inhibited from further extension when they touch another nerve fibre. Dickinson (1964a, b) showed that the germ-tubes of the rust fungus *Puccinia coronata* grew at *right angles* to visible lines in various artificial membranes, and suggested that the alignment of a fibrillar component of the cell wall was involved in this orientation, which he termed thigmotropism. Maheshwari and Hildebrandt (1967) found that the germ-tubes of *Puccinia antirrhini* grew at right angles both to the cuticular ridges on the leaves of its host *Antirrhinum majus* and to the ridges on collodion or cellulose acetate replicas of the leaf surface.

Contact guidance also takes the form of cells migrating along slime tracks left by other cells. Such tracks are likely to consist in part of oriented macromolecules. Another form of contact guidance occurs in aggregation of the cellular slime mould *Dictyostelium discoideum*, the front end of a migrating amoeba tending to adhere to the rear end of the one in front, resulting in neat lines of aggregating cells (Shaffer, 1964).

The various forms of contact guidance are clearly of considerable importance in oriented movement. However, it seems misleading to call them taxes or tropisms (thigmo- or elastico-). Contact guidance imposes restraints on the direction in which a cell can grow or move, but does not in itself determine which direction is chosen. An analogy is a railway track; it severely restricts the directions in which a train will move, but does not settle which of the two possible directions will be chosen—other factors determine this. Similarly, it would seem that in contact guidance the actual direction of movement must be determined by an existing polarity or another tactic or tropic agent.

Carter (1965, 1967) showed that fibroblasts will migrate up an adhesion gradient, a phenomenon he called *haptotaxis*. He suggested that *haptotaxis* was of great importance *in vivo* and could account for many phenomena. He pointed out that a local release of a substance that modified adhesiveness could give rise to a gradient of adhesiveness in a tissue and could lead to the migration of motile cells towards or away from the point of release, thus giving an apparent chemotaxis. Carter also demonstrated how haptotaxis could provide a basis for contact guidance, since he showed that movements of cells can be restricted to a long narrow region of high adhesion surrounded by areas of low adhesion. Harris (1973) concluded that haptotaxis was due to random extension of pseudopodia, with those in regions of low adhesiveness breaking contact with the substratum more readily than those in regions of high adhesiveness, thus resulting in the cell tending to move up any adhesion gradient.

F. Heat

Bonner *et al.* (1950) showed that pseudoplasmodia ("slugs") of the cellular slime mould *Dictyostelium discoideum* showed positive *thermotaxis* (travelled up a temperature gradient). The experiments were carried out at 24°C and orientation was obtained in gradients of 0·05°C/cm but not at 0·02°C/cm. The response is a very sensitive one, as 0·05°C/cm represents a temperature difference between the two sides of the pseudoplasmodium of 0·0005°C.

G. Electricity

Oriented movement in response to a weak electric current, *galvanotaxis* or *electrotaxis*, is widespread and distinct from the elctrophoretic movement of small charged particles such as bacteria. It occurs in ciliates, including *Paramecium* (Eckert, 1972), Myxomycete plasmodia (Anderson, 1951) and fungal zoospores (Troutman and Wills, 1964; Katsura and Miyata, 1971). Migration is normally towards the cathode. In Bracken spermatozoa Brokaw (1958b) showed galvanotaxis occurred in the presence of malate, migration being towards the anode. In *Phytophthora capsici* migration of zoospores can occur towards cathode or anode depending on which organic acids are present (Katsura and Miyata, 1971). By applying different current intensities for different times Masson *et al.* (1952) used the galvanotactic response for fractionating rumen ciliates.

III. The nature of the response

At the earliest phase in research on a taxis or tropism all that is known is that something is proving attractive or repellent. Subsequently progress is made in clarifying the initial and final phases of the effect—precisely what is the nature of the stimulus and how is movement affected? Progress in understanding the sequence of events between stimulus and response can then begin either from the stimulus (e.g. what is the nature of the chemoreceptor?) or from the response (e.g. what changes in flagellum movement produce the observed changes in the path of the organism?). In a few instances one can now envisage a plausible sequence of events from stimulus through reception to the final response, although in all cases areas of mystery remain.

In some organisms the mechanism of movement is so ill-understood as to prevent any useful discussion of the way it is influenced in taxis, as for example in Myxobacteria. With all micro-organisms and cells a better understanding of growth and movement is needed, but in some instances, considered below, enough is known for useful discussion.

A. Tropisms of fungi

In most fungal hyphae (including germ-tubes, but excluding some spectacular sporangiophores and conidiophores—see below) growth is confined to the hemispherical apex of the hypha, and is thought to be maximal at the very tip (Bartnicki-Garcia, 1973). Hence if the area of maximal extension is shifted a little to one side, then growth towards that side will occur, and changed orientation result. So positive tropic agents will act by stimulating wall extension. This (Bartnicki-Garcia, 1973) involves both a plasticising or partial lysis of the wall and the synthesis of further wall material. Although lysis and synthesis occur *in situ*, materials for the synthetic process and the lytic and synthetic enzymes have to be transported to the wall, the enzymes in vesicles. In the fungal hypha protoplasmic streaming is a major factor in translocation of materials to the growing point. Asymmetry in growth and hence tropism could therefore be produced by two processes—changed enzyme activity at the wall or an asymmetry in the protoplasmic streaming that conveys materials to the wall. It would seem likely that both mechanisms occur. Thus a gravitational displacement of a central vacuole (section II D) will restrict supply of enzymes or materials, whereas light might act on enzymes in the plasmalemma.

Sporangiophores and some conidiophores have an intercalary growth zone. With these, positive tropism will result from a reduced growth rate on the near side, with relatively greater growth on the far side causing the sporangiophore to swing towards the near side. Hence a positive tropic effect will normally be obtained through a depression of the rate of near-side wall extension. An exception is in phototropism where light stimulates growth, but on the far side through a lens effect.

One fungal tropic response, the phototropism of the *Phycomyces* sporangiophore, is perhaps the most thoroughly analysed of all tropisms, particular attention having been paid to sensory adaptation (Bergman *et al.*, 1969). Such adaptation is essential in any sensory system that has to respond with equal efficiency to a small percentage difference in a very weak stimulus (i.e. dim light) or a very strong stimulus (i.e. full daylight).

The sporangiophore responds also to gravity (section IID), stretch (IIE) and to solid objects by avoidance, probably a negative hydrotropism (IIC). Progress has been made through the use of mutants in outlining the pathways by which these different sensory imputs influence the common output, curvature based on differential growth (Bergman, Eslava and Cerdá-Olmedo, 1973).

B. Amoeboid and plasmodial taxes

The pseudoplasmodium ("slug") of the cellular slime mould *Dictyostelium discoideum* consists of about 10^5 amoebae enclosed in a slime sheath. The sheath is plastic at the front tip of the "slug" but coagulates further back; this maintains the polarity of the migrating "slug", only permitting it to move forward (Loomis, 1972). The slugs are positively phototactic (Poff, Butler and Loomis, 1973). A lens effect operates, light exerting its maximum effect on the side of the slug away from the source, but bringing about the turning of the tip of the "slug" towards the source (Francis, 1964). It was suggested that light did this by accelerating the hardening of the slime sheath on the far side of the "slug". Thermotaxis (section IIF) might operate through heat stimulating the activity of the amoebae or delaying sheath hardening on the warmer side of the apex. The apex of the "slug" and the slime sheath are analogous to the hyphal apex and wall in fungi, and the taxic mechanism of cellular slime moulds could be similar in principle to that of the tropisms of fungi.

The Myxomycete plasmodium is a mass of protoplasm which is not sub-divided into cells and which may cover many cm^2. Within the protoplasm there are channels in which protoplasmic streaming occurs at rates of up to 1 mm/sec, with reversals in the direction of flow occurring at about 1 min intervals. This shuttle streaming is the consequence of rhythmic contractions occurring at different times in different parts of the plasmodium—see Kamiya (1959) for a detailed review and Korohoda, Rakoczy and Walczak (1970) for recent references. If the volume of protoplasm transported in one direction is greater than that transported in the opposite direction, then movement of the plasmodium will occur (Jahn and Bovee, 1964). The net direction of transport will be determined by the relative vigour of contractions in different regions and the resistance offered in other areas, presumably including that due to the slime coat (glycocalyx) which surrounds the plasmodium. Many types of taxis have been reported in Myxomycete plasmodia, including phototaxis (Rakoczy, 1963) and chemotaxis to nutrients (Carlile, 1970b). A positive tactic agent may operate by reducing contractile activity thus reducing flow away from the area most influenced, a negative by enhancing contractile activity.

The mechanism of locomotion of amoeboid cells remains highly controversial and there may well be major differences in mechanism between the very diverse free-living amoebae and the amoeboid cells of mammalian tissues. As mentioned earlier, fibroblasts undergoing haptotaxis produce pseudopodia at random but those in contact with regions of high adhesion are more persistent (Harris, 1973). Leucocytes undergoing chemotaxis produce pseudopodia at random but the contents of the rest of the cell tend to flow preferentially into the pseudopodium nearest the source of the attractant (Ramsey, 1972). In *Amoeba proteus*, however, Jeon and Bell (1965) reported preferential *production* of pseudopods on the side nearest an attractant.

Tentatively, it is suggested that the chemotaxis of amoeboid cells and the tropism of fungal hyphae have some resemblance—both may involve changes in the magnitude and direction of protoplasmic streaming and in the local resistance offered by a "cytoskeleton" consisting of glycocalyx (cell wall or slime sheath) and perhaps cortical fibrils.

C. *Flagellate and ciliate taxes*

1. *Taxes of bacteria*

Fraenkel and Gunn (1961) explained how klinokinesis with adaptation could permit an organism to respond appropriately to attractants or repellents, and cited the behaviour of a flatworm as their example. Since zoologists no longer regard this interpretation of the flatworm's behaviour as valid (see preface to Loeb, 1973), it is fortunate that the tactic responses of bacteria now provide an excellent example of the phenomenon. Many microbiologists have used the term *phobotaxis*, since what is often observed is a "shock reaction" resulting in a reversal or a change in direction when the organism is travelling the "wrong" way, with such changes in direction being infrequent when the organism is going the "right" way. Some classical accounts of phobotaxis concern the response of photosynthetic bacteria when they pass from an illuminated field into darkness (Chapter 2) but the most intensive recent work has been with the positive chemotaxis of bacteria to amino acids, reviewed by Koshland (1974) and in Chapter 3.

Bacterial cells in isotropic solutions undergo frequent spontaneous and random changes in their direction of swimming. McNab and Koshland (1972) working with *Salmonella typhimurium* showed that a sudden increase in the concentrations of an attractant amino-acid *temporarily* suppressed the frequency of these directional changes and that a sudden decrease enhanced their frequency. Thus a bacterium travelling down an attractant gradient would soon change direction through "tumbling" whereas one going up the gradient would continue for a long time in a straight path. Berg and Brown (1972), by tracking

Escherichia coli cells in isotropic solutions and in gradients of an attractant, observed that this occurred. A turn is probably the result of a "tumble" due to temporarily unco-ordinated flagellum activity whereas long smooth runs result from the flagella operating as a single unit. Tsang, McNab and Koshland (1973) found that tumbles occur less frequently when bacteria *descend* a repellent gradient as well as when they ascend an attractant gradient. It was concluded that repellents and attractants, although having different specific receptors, act through a common effector system.

Berg and Anderson (1973) and Silverman and Simon (1974) suggested that bacteria move by the rotation of semi-rigid flagella. Larsen *et al.* (1974) found that in the presence of an attractant counter-clockwise flagellar rotation occurred and with repellents clockwise rotation. It was concluded that counter-clockwise rotation resulted in long smooth runs and clockwise rotation in abrupt direction changes or "tumbles" through unco-ordinated flagellum activity.

The earliest stage of the chemotatic response of *E. coli*, the binding of the attractant by a specific protein receptor located in or near the cell membrane, is considered further in Chapter 3. The intervening steps that occur between a change in amount of attractant bound to a chemo-receptor and changed flagellar activity remain unknown, although changes in the electrical properties of the cell membrane could be involved, as speculated by Doetsch (1972).

An intriguing observation is that chemotaxis of *Pseudomonas fluorescens* to nutrient broth is enhanced by the stimulants amphetamine and epinephrine. (Chet, Henis and Mitchell, 1973).

2. *Taxes of eukaryotic flagellate cells*

True taxis, in which steering towards the source of an attractant occurs, is widespread in flagellate eukaryotic cells, as demonstrated by tracking experiments with fungal zoospores (Royle and Hickman, 1964; Allen and Newhook, 1973) and fern spermatozoa (Brokaw, 1958b, 1974). Klinokinesis also occurs in eukaryotic cells, but has attracted rather less attention than true taxis. It may occur when a concentration gradient is not steep enough for a spatial difference in intensity to be detected, hence an organism may approach the source of an attractant initially through klinokinesis, and later, as the gradient of the attractant becomes steeper, through a true taxis. Allen and Newhook (1974) have shown that ethanol, which causes true positive taxis with *Phytophthora cinnamomi* zoospores, will suppress spontaneous turning of the zoospores, which suggests that it could also act by klinokinesis. Klinokinesis may also suffice as a response towards repellents, for which a precisely oriented response is not needed, a phobotactic reaction to a critical concentration (Allen and Harvey, 1974) being sufficient.

A true taxis will normally imply the existence of two or more receptors at different points on the cell surface so that *simultaneous* comparison of intensities can ascertain the direction of a gradient. However, in positive phototaxis steering can be achieved by means of the response of a single receptor responding to shading by the cell body, re-orientation occurring until the receptor, located at the front of the cell, receives maximum stimulation. In *Euglena,* which rotates as it swims, turning occurs so as to eliminate the periodic shading of the receptor near the flagellum base by a pigment patch, the "eyespot" (Chapter 2). Since the basis of positive phototaxis is the comparison of intensities at a single receptor at successive times, it has features in common with the temporal comparison of intensities in klinokinesis (phobotaxis).

Recently some studies have been carried out on the changes in flagellum activity which result in the steering of the cell. Miles and Holwill (1969) showed that asymmetrical movement of the single flagellum was responsible for apparently spontaneous turning in the zoospores of the watermould *Blastocladiella emersoni* and hence this was presumably the basis for the chemotactic responses which occur in this species (Carlile, unpublished). Miller and Brokaw (1970) found that asymmetrical flagellum bending was responsible for chemotactic turns in the uniflagellate spermatozoa of the coelenterate *Tubularia.* It seems that individual spermatozoa in this species can turn in one direction only, and that depending on the direction of the source of the attractant, a simple turn or a series of loops may be needed. It was postulated that the more precise turning of Bracken spermatozoa was a consequence of the latter having many flagella capable of being influenced by a tactic agent, an averaging out of effects resulting in a smooth response rather than the drastic one of uniflagellate forms. Fraenkel and Gunn distinguish between klinotaxis, in which deviations from a direct path towards an attractant occur, followed by turning away from the side on which the attractant concentration is less (a form of behaviour also referred to as "hunting"), and topotaxis, in which deviation does not occur. Perhaps the behaviour of uniflagellate cells approximates to klinotaxis and multiflagellate cells to topotaxis.

There is, as yet, little information about the receptors involved in eukaryotic taxes, although in chemotaxis highly specific receptors are likely to be involved, as for example with the very specific response of *Allomyces* gametes to *l*-sirenin (Machlis, 1973a). Little is known about the way in which the tactic signal is transmitted from receptor to flagellum, although clues are provided by work on the contact reaction in ciliates (see below) and the very widespread requirement for calcium (or tactic or tropic activity of calcium) in eukaryote taxes and tropisms as in *Allomyces* gametes (Machlis, 1973b), pollen tubes (Mascarenhas and Machlis, 1962b, 1964), and Bracken spermatozoa (Brokaw, 1974).

3. *Ciliate taxes*

Ciliate taxes have on the whole been badly neglected, the rather meagre knowledge available having been reviewed by Dryl (1973). However, studies on the contact reaction in *Paramecium* have provided perhaps the most fully understood sequence of events between a stimulus and a response for any micro-organism (Eckert, 1972). When Paramecium collides with an obstacle the cell membrane is stretched at the anterior end leading to a local increase in membrane conductance and an inward receptor current through the stimulated membrane. The electrotonic spread of the receptor current produces an outward current through the rest of the membrane, depolarisation, increased calcium conductance and an influx of calcium ions. The increased intracellular calcium concentration leads to the effective stroke in the beating of the cilia being forwards instead of backwards. This reversal of the normal direction of ciliary beat results in the cells swimming backwards. Subsequently calcium ions are pumped out of the cell and the normal direction of ciliary beat and cell movement resumed. The role of the internal calcium concentration in ciliary reversal was confirmed by Naitoh and Kaneko (1972) who found that triton-extracted *Paramecia* placed in an appropriate solution were capable of swimming and that at calcium concentrations above 10^{-6} M swam in reverse due to reversal in the direction of ciliary beat. Studies on the role of membrane depolarisation in the transmission of the signal from the receptor area to cilia or flagella and on the role of calcium in modulating flagellum activity are needed for other taxes and other organisms. The control of ciliary activity in Protozoa has been reviewed by Naitoh and Eckert (1974).

IV. Biological significance of taxes and tropisms

It is easy to speculate on the possible role of a taxis or tropism in the life of an organism, but less easy to be certain about the precise significance of a particular tactic or tropic response. For example, the germ-tubes of a fungus may show oriented growth towards and into the stomata of plants, but it will be uncertain whether a positive tropism to water, carbon dioxide or nutrients or a negative phototropism is responsible. Proof of the ability of the fungus to show some or all of these responses will still not establish which is the most significant in nature—for this it will be necessary to demonstrate the occurrence at the stomata of gradients of appropriate magnitude to account for the response. The positive phototaxis of a photosynthetic alga at low light intensities is readily attributable to the need to reach light sufficient for photosynthesis, and negative phototaxis at high intensities to the need to avoid damagingly intense light; it would, however, be good to know whether the zone of reversal from positive to negative phototaxis corresponds to the optimal light intensity for growth in ecologically realistic conditions. Some taxes and tropisms may be without significance in nature, being a fortuitous consequence of the biochemical processes necessary for response to some other agent. Thus electrotaxis may perhaps be an accidental consequence of a chemotactic sensitivity to various ions. However, caution is required in reaching negative conclusions about the utility of this or any other biological activity. Weak electric currents do occur in nature, and Troutman and Wills (1964) have postulated that electrotaxis is involved in attracting fungal zoospores to plant roots. Much more work is needed on the physiological ecology of taxes and tropisms, so in this section it is possible to do little more than discuss the probable role of a few of the better known oriented responses.

A. Selection of environment

The probable role of positive and negative phototaxis in placing photosynthetic micro-organisms at light intensities optimal for growth has been commented upon above. Negative phototropism occurs in a wide range of micro-organisms and cells in which penetration of the substratum either for anchorage or host invasion is required (Haupt, 1965). It is fairly common in the germination of plant pathogenic fungi (Carlile, 1966), algae such as the brown sea-weed, *Fucus* (Jaffe, 1958), mosses (Jaffe and Etzold, 1965) and ferns (Jaffe and Etzold, 1962).

Positive chemotactic responses to oxygen, amino-acids and sugars are widespread in bacteria. Barrachini and Sherris (1959) demonstrated positive aerotaxis (chemotaxis to oxygen) in 23 of the 24 aerobic and facultatively anaerobic motile bacteria that they examined. Taxes by a wide range of bacteria to amino-acids and sugars were shown by Seymour and Doetsch (1973), who concluded that the response was unlikely to be of any biological significance since species varied greatly in respect of which amino-acids and sugars produced a response. Such a result, however, was to be expected, as bacteria vary widely in their qualitative and quantitative nutrient requirements, and furthermore, since a standard procedure was used throughout, conditions would have been optimal for only a small proportion of the species tested. Chet, Fogel and Mitchell (1971) showed that a *Pseudomonas* capable of lysing the fungus *Pythium debaryanum* showed taxis to *Pythium*, and one of lysing the alga *Skeletonema costatum*, taxis to *Skeletonema*. *Pseudomonas* strains incapable of lysing either showed only a weak response. Bell and Mitchell (1972) showed that filtrates from ageing cultures of *Skeletonema* caused positive chemotaxis in, and stimulated growth of, several marine bacteria. Mitchell, Fogel and Chet (1972) found that hydrocarbons, including crude oil, inhibited the chemotactic responses of some marine bacteria and considered the ecological consequences of this effect. The chemotaxis of the plant pathogen *Pseudonomas lachrymans* to host plant extracts and exudates was studied by Chet, Zilberstein and Henis (1973). An infection process of great ecological significance which could well involve chemotaxis is the invasion of legume root hairs by *Rhizobium*, essential for the development of nitrogen-fixing root nodules, although this possibility does not seem to have been investigated.

As well as positive chemotaxis to favourable conditions, negative chemotaxis to unfavourable conditions occurs in bacteria. Barrachini and Sherris (1959) found negative aerotaxis in 3 of the 4 members of the obligately anaerobic bacterial genus *Clostridium*. Negative taxis to unfavourable pH was shown to have a survival value in *Pseudomas fluorescens* by Smith and Doetsch (1969). Negative taxes to poisons were shown in *E. coli* by Tso and Adler (1974) and in marine bacteria by Young and Mitchell (1973).

Taxis to nutrients in eukaryotic micro-organisms is widespread. Many amoeboid cells show chemotaxis to bacteria (Chapter 4) and the zoospores of the lower fungi to amino-acids, sugars or exudates from plant roots (Chapter 5). Nutrient chemotropism in fungi seems rare, except perhaps a positive aerotropism (Robinson, 1973). However, the germ-tubes of the mycoparasitic fungi *Piptocephalis virginiana* and *Dispira cornuta* show strong positive chemotropism to the hyphae of host fungi, and two other mycoparasites, *Calcarisporium parasiticum* and *Gonatobotrys simplex* secret an attractant from their germinating spores which induces positive tropism by host species (Barnett and Binder, 1973).

B. *Morphogenesis, repair and defense*

Much of the morphogenetic activity of micro-organisms is concerned with the production of fruiting bodies from which spores are dispersed, often after genetic recombination has occurred. So the distinction between oriented responses which are involved in morphogenesis and those which are involved in genetic recombination and dispersal (sections IV C and D) will be rather arbitrary, as there are many which could well be assigned to either category.

The most spectacular display of the role of chemotaxis in microbial morphogenesis is seen in the aggregation of the amoebae of the Cellular Slime Moulds to form a pseudoplasmodium (Chapter 4). Contact guidance is also involved. Chemotropism and contact guidance are probably also responsible for the less thoroughly studied aggregation of Myxobacteria to form a fruiting body (p. 5). In fungi an autotropism is responsible (Chapter 5) for the hyphal anastomoses which convert colonies of higher fungi from a radiating set of hyphae to a three-dimensional network. In coremium formation in some *Penicillium* spp, phototropism is a factor in achieving the parallel alignment of hyphae needed for the unbranched coremia produced in light (Carlile, Dickens and Schipper, 1962). As the morphogenetic processes involved in fruiting body formation are more fully elucidated, further examples of the role of sexual chemotropism, autotropism and phototropism are likely to emerge.

A special form of morphogenetic activity is the repair of damage. In the red alga *Griffithsia pacifica* Waaland and Cleland (1974) have shown that the death of an intercalary cell in a filament is followed by projections from the cells above and below growing through the dead cell towards each other, meeting, fusing and expanding to replace the dead cell. Their photographs convincingly demonstrate attraction between the new growths.

The possible role of cellular taxes in the morphogenesis of higher organisms is a controversial topic. Edelstein (1971) has proposed a chemotactic theory of cell sorting but other models seem to be more popular at present. Crick (1970) has, however, shown that diffusion could be responsible for establishing morphogenetic gradients of the magnitude postulated by embryologists (Wolpert, 1969). So it would seem that situations exist in development where chemotaxis could have a role. Haptotaxis and contact guidance, both demonstrated in mammalian cells (section II E), could also have a role. Although it is as well to be sceptical of any sweeping chemotactic theories of embryogenesis, it is likely that various roles for oriented cellular movement in the morphogenesis of higher organisms will emerge.

A major role of chemotaxis in higher organisms is in defence, leucocytes moving towards invading bacteria and destroying them (Chapter 6). Leucocytes will also show chemotaxis to dead cells, an effect termed necrotaxis by Bessis (1964).

C. *Genetic recombination*

An attractant released by the female gamete and attracting the male is one method of increasing the chances of fertilization, and hence bringing about genetic recombination. Chemotaxis of male gametes to the female gamete or gametangium is widespread in the Algae and Lower Fungi, as also is chemotropic guidance in species lacking mobile gametes (Chapter 5). Several of the attractants have been isolated, and are relatively complex molecules and most would appear to be highly specific, acting only on the gametes of one or of a few closely related species.

Chemotaxis of male gametes to the female is also firmly established in many lower plants (Machlis and Rawitscher-Kunkel, 1963) although usually relatively simple attractants such as organic acids are involved

(Rothschild, 1956). In the liverwort *Sphaerocarpus donnelli*, however, the attractant is of molecular weight ca. 2,000–4,000 (Scheider, 1968). In the most thoroughly studied species, the Bracken fern, both calcium and malic acid are chemotactically active, and both must be present for chemotaxis to occur (Brokaw, 1974). In at least some higher plants chemotropism assists the pollen tube in penetrating the female tissues, and in some calcium is chemotropically effective and in some not (Mascarenhas, 1973). It is difficult to envisage a chemical gradient of length sufficient to guide the pollen tube all the way from stigma to ovum, and presumably for the greater part of the journey down the style the pollen tube is following the line of least resistance.

Chemotaxis of male gametes has been firmly established in one group of lower animals, the Coelenterates (Miller, 1973); whether it occurs in higher animals remains controversial. Bretherton and Rothschild (1961) have demonstrated positive rheotaxis (up-stream swimming) in bull and human spermatozoa, which presumably assists the spermatozoa in travelling up the oviduct.

D. Dispersal

The spores of terrestrial fungi are usually dispersed through the air. Hence the spore-bearing structures (sporophores or fruiting bodies) must, as they grow, be manoeuvred past obstacles and out into the light and air, where spores can be scattered by air currents or rain splash. In many fungi spores are actively discharged, and it is important that they should be shot into the open air where currents can disperse them further rather than into nearby obstacles. With the above considerations in mind, it is easy to see why positive phototropism is very common in the fruiting bodies of fungi (Carlile, 1965), and many examples of its role in dispersal are provided by Ingold (1953).

Less widespread with fungal sporing structures is negative geotropism, which at night or in dark situations will keep a sporophore or fruiting body growing in the right direction. The negative geotropic response, where it occurs, is commonly suppressed by light at quite low intensities, as in *Phycomyces* (Bergmann *et al.*, 1969). This would seem biologically advantageous, since light when present is likely to be a better guide than gravity. In *Phycomyces* an avoidance response (Johnson and Gamov, 1971), possibly a negative hydrotropism, also assists in avoiding obstacles. Some negatively chemotropic gas, but probably not water vapour, has a role in the orientation and spacing out of the sporophores of the cellular slime mould *Dictyostelium* (Bonner and Dodd, 1962).

Positive phototaxis of the zoospores of some aquatic fungi (Kazama, 1972; Robertson, 1972) presumably has a role in dispersal, getting the zoospores away from the zoosporangia and into open water. The positive phototaxis of the pseudoplasmodium of the Acrasiales (section III B) and the plasmodium of some Myxomycetes also will tend to result in fruit body formation in exposed locations suitable for spore dispersal.

V. Evolution of sensory and communication systems

A. *Receptors*

Phototaxis and chemotaxis are widespread in both prokaryotes and eukaryotes. There is little evidence for taxis to other agents in prokaryotes, or on the nature of the receptors for these agents in eukaryotic micro-organisms. Discussion will therefore be limited to photoreceptors and chemoreceptors.

1. *Photoreceptors*

Phototaxis is a response which is very widespread in photosynthetic organisms and is of obvious value to them, but it is fairly uncommon among other cells. Therefore it probably first arose in prokaryotes already capable of photosynthesis and having chlorophyll and accessory photo-synthetic pigments in their cell membranes and membrane invaginations. Hence these pigments were already available for the role of receptor pigments in phototaxis, and retain this role in all photosynthetic bacteria and blue-green algae in which the phototactic action spectrum has been adequately studied (Chapter 2).

According to current views on evolution (Margulis, 1970), photo-synthesis in eukaryotes is thought to have originated by primitive eukaryotic species ingesting photosynthetic prokaryotes and retaining them to give at first photosynthetic endosymbionts, and later chloroplasts, with the pigments characteristic of the various algal groups (Stanier, 1974). There is no obvious way in which these endosymbionts could swiftly influence the movement of their eukaryote hosts owing to their location deep in the cell and away from the host's locomotory organelles. Hence the host, unless already phototactic, which is not very probable, would have had to evolve its own phototactic system and receptors so as to place itself in optimal light intensities for exploiting its endosymbionts. Thus eukaryote phototaxis probably evolved separately in each algal group as it acquired endosymbionts and also in some non-photosynthetic forms that show phototaxis or phototropism. An intriguing possibility is that the phototactic sensitivity of the captured algae survives today as the capacity of chloroplasts to orient to light. However, this seems unlikely, as light-oriented chloroplast movement seems to have a flavin photoreceptor (Zurzycki, 1972) and to be passive (i.e. moved by the cytoplasm) rather than active (Haupt and Schönbohm, 1970).

Action spectra for phototaxis in eukaryotes are commonly consistent with a flavin photoreceptor and sometimes also a haem receptor. Such pigments would already occur in aerobic cells, both prokaryote and eukaryote, as part of the respiratory electron transfer system, and hence be available as potential photoreceptors. McNab and Koshland (1974) have shown that high intensity light will induce "tumbling", an essential part of the chemotactic response, in *Salmonella typhimurium*, and have obtained a crude action spectrum suggesting a flavin photoreceptor. As far as is known *Salmonella* is incapable of phototxis, but it would seem probable that under appropriate selection increased light sensitivity and a phototactic response could evolve. The respiratory pigments of prokaryotes are located in the cell membrane and hence are well placed to influence photosensitivity. Further studies on the photoresponses, especially action spectra, of the few non-photosynthetic prokaryotes known to show phototaxis, such as the myxobacterium found by Aschner and Chorin-Kirsch (1970) to show negative phototaxis as swarms but not as individuals, would be of interest.

In *Euglena* the site of maximum photosensitivity is a swelling near the base of the flagellum (Chapter 2). This association of flagellum and photoreceptor is of great evolutionary interest as it would seem that from it or something similar the light sensitive cells of higher animals may have evolved (Leedale, 1967). The retinal cells of vertebrates retain the ring of 9 fibrils characteristic of eukaryote flagella (Barber, 1974).

2. *Chemoreceptors*

In *Escherichia coli* the same binding protein in the cell membrane serves for the initial step in galactose uptake and chemotaxis to galactose (Chapter 3). So it is probable that chemoreceptor mechanisms can evolve from pre-existing transport mechanisms for nutrients such as amino-acids, sugars and ions, and that in *E. coli* each of the receptors for positive chemotaxis has arisen in this way. How receptors for negative chemotaxis arose is less clear, but again binding proteins in the cell membrane were presumably involved. Chemosensitivity to various nutrients has been acquired or lost many times in the evolution of different groups of micro-organisms. Receptors for nutrients are in general of relatively low sensitivity, responding to concentrations higher than about 10^{-6} M, and of fairly low specificity, giving some response to several related compounds. The receptors for the fungal sex-hormones (Gooday, 1973) are of very high sensitivity, responding to concentrations of as low as 10^{-12} M, and probably of very high specificity—*Allomyces* male gametes, for example, have not been found to show a response to any chemical other than *l*-sirenin (Machlis, 1973). A curious correlation is that many cells capable of very sensitive chemotactic or chemotropic responses, such as the male gametes of *Allomyces* or the zygophores of *Mucor*, contain high carotenoid concentrations, and that this also occurs in the olfactory organs of higher animals where it has been thought to have a role in chemoreception (Rosenberg, Misra and Switzer, 1968).

B. *Signal transduction and transmission*

Phototaxis in the photosynthetic prokaryotes could involve disturbances in electron flow in the photosynthetic electron transport system (Chapter 2). In non-photosynthetic micro-organisms a disturbance in the electron flow in respiratory electron transport is feasible. Perturbation of the electron flow in the electron transport system in *Salmonella typhimurium* can generate "tumbling" and Doetsch (1972) has suggested that all bacterial sensory responses involve changes in membrane potential. In *Paramecium* detailed studies have been carried out on the role of changes in membrane potential in controlling ciliary activity (Eckert, 1972). So although the initial steps in signal transduction will vary depending on the nature of the stimulus, and the final steps depending on the nature of the effector organelle, the transmission of the stimulus from sensor to effector along the cell membrane may be similar for many responses in a wide range of organisms. Such signal transmission is of great interest as the precursor of the processes responsible for signal amplification in vertebrate sensory cells, such as those of the retina (Delbrück, 1972), and transmission in the neurones.

C. *Effector mechanisms*

There are several very different forms of motility in micro-organisms, such as amoeboid movement, the gliding of myxobacteria and blue-green algae and swimming by means of flagella. The mechanisms of locomotion involved are so different that it is reasonable to suppose that they have arisen independently and therefore that both movement and oriented movement (taxis), as well as oriented growth (tropism), have come into existence several times.

Taxes must have been preceded by active but random motility. Such motility would have been advantageous, as it would bring about more effective dispersal than Brownian movement (for very small organisms) or chance currents (for larger ones). The dispersal of *E. coli* by random motility has been studied by Adler and Dahl (1967). As soon as active but random motility existed, the evolution of non-random motility, through the development of receptors responding to important environmental variables, became possible. Probably the earliest such responses were simple shock reactions when some obnoxious factor reached a critical level. However, adaptation in a sensory system has the great advantages of achieving sensitivity over a very wide range of stimulus intensities (Delbrück and Reichardt, 1956) and terminating a response that is wasting energy and failing to achieve improved conditions. Hence an early evolution of sensory adaptation could be expected, making possible the klinokinesis with adaptation (phobotaxis) which is widespread in bacteria today. The value of such a response in the selection of environment is so great that it is improbable that organisms capable only of active but random movement would persist for long after the evolution of related strains capable of taxis.

True taxis, in which oriented movement of the cell occurs, is likely to have evolved later than phototaxis, and may well be absent in prokaryotes. The topotaxis of blue-green algae is a rather special taxis involving not steering but an effect of light on the duration of the autonomous periodic reversals in direction of movement of a filament already oriented by chance. The possibility that a true taxis occurs in the aggregation of myxobacteria to form a fruiting body deserves investigation.

The main reason for the rarity or absence of true taxes in prokaryotes would seem to be the small size of the cells in most species. Topotaxis requires that an organism should be able to detect different stimulus intensities at different points on the cell surface, and for cells of bacterial dimensions such differences would commonly be so slight as to be undetectable. Phobotaxis, on the other hand, can involve the comparison of intensities before and after 10 seconds travelling, permitting, in the case of *Salmonella* (Koshland, 1974), sampling of a gradient at points 300 μm instead of 2 μm, the length of the cell, apart. This requires an analytical accuracy of 1 in 10^2, which is feasible, instead of 1 in 10^4, which is not. So phobotaxis would seem the only practical form of taxis with most prokaryotes. For larger cells topotaxis is feasible and will permit a swifter and more accurate response. Tropisms do not seem to have been reported from prokaryotes (except for the rather special case of elasticotropism, involving contact guidance) and again it would seem rather unlikely as the small size of cells would not permit a perceptible gradient of an environmental factor across the cell.

The photophobotaxis of *Euglena* would seem to represent an intermediate stage between phobotaxis and topotaxis; orientation takes place as a result of a series of phobotactic responses occurring when the receptor near the flagellum base is periodically shadowed (Chapter 2). Phobotaxis remains widespread in eukaryotic micro-organisms showing topotactic responses, and is presumably of value when gradients are not steep enough for topotaxis to operate.

D. Intercellular communication

Once an organism possessed a sensory system responding to chemicals, the evolution of a communication system became possible. A response to a chemical in the environment could become a response to the same chemical fortuitously emitted by another species and finally by cells of the same species. The final step is illustrated by the amoebae of *Dictyostelium* which, during their growth, respond by chemotaxis to cyclic AMP emitted by the bacteria on which they feed; this sensitivity has been exploited in the aggregation phase when amoebae respond by chemotaxis to cyclic AMP emitted by "founder" amoebae, and then, after a delay, themselves emit cyclic AMP attracting further amoebae and creating a relay system (Chapter 4). Thus the presence of a sensory system has permitted the evolution of a communication system with a role in morphogenesis.

Chemical communication is also seen in the algae and fungi in which male gametes respond to a hormone released by the female (Chapter 5). The climax of chemical communication in micro-organisms occurs in those fungi in which a series of hormonal signals are exchanged between strains of different sexes or mating types (Gooday, 1973). For example, in the water mould *Achlya*, antheridiol (hormone A) from female plants stimulates the differentiation of reproductive structures in the male which then release hormone B which brings about further differentiation in the female.

The existence of chemical communication between cells must have been a prerequisite for the evolution of multicellular organisms. In plants, intercellular communication remains largely hormonal. In animals hormonal communication has been supplemented by neuronal. Even here, however, although the transmission of signals along a neurone is electrical, the communication across synapses remains chemical. Sensory systems permitted the evolution of intercellular communication, which in turn made possible the evolution of the more sophisticated sensory systems of higher organisms, with ranges of highly specialized receptor cells of great sensitivity and nervous systems capable of evaluating information and delaying responses. Hence in evolution the development of sensory and communication systems have been closely related.

There is now much information on the sensory and communication systems of micro-organisms and the most complex animals (insects and vertebrates) and plants. There is, however, a relative lack of sophisticated work on the sensory responses and hormonal systems of plants and of animals with relatively simple nervous systems. The nematodes, for example, display a remarkable range of sensory responses (Croll, 1970) including chemotaxis to amino-acids, salts and cyclic AMP (Ward, 1973). Further studies in this area would illuminate an important and little understood aspect of evolution.

I wish to thank Professor J. S. Kennedy, F.R.S. and Dr Graham Gooday for reading the manuscript and suggesting improvements.

References

ADLER, J. and DAHL, M. M. (1967). *J. Gen. Microbiol.* **46,** 161–173.
ADLER, J. and TSO, W. W. (1974). *Science* **184,** 1292–1294.
ALLEN, R. N. and HARVEY, J. D. (1974). *J. Gen. Microbiol.* **84,** 28–38.
ALLEN, R. N. and NEWHOOK, F. J. (1973). *Trans. Brit. Mycol. Soc.* **61,** 287–302.
ALLEN, R. N. and NEWHOOK, F. J. (1974). *Trans. Brit. Mycol. Soc.* **63,** 383–385.
ANDERSON, J. D. (1951). *J. Gen. Physiol.* **35,** 1–16.
ASCHNER, M. and CHORIN-KIRSCH, I. (1970). *Arch. Mikrobiol.* **74,** 308–314.
BANBURY, G. H. (1962). *In* "Encyclopedia of Plant Physiology" (W. Ruhland, ed.), Vol. 17, part 2, pp. 344–377, Springer, Berlin.
BARBER, V. C. (1974). *In* "Cilia and Flagella" (M. A. Sleigh, ed.), pp. 403–433, Academic Press, London.
BARNETT, H. L. and BINDER, F. L. (1973). *Ann. Rev. Phytopath.* **11,** 273–292.
BARRACHINI, O. and SHERRIS, J. C. (1959). *J. Path. Bact.* **77,** 565–574.
BARTNICKI-GARCIA, S. (1973). *In* "Microbial Differentiation", 23rd Symposium of the Society for General Microbiology (J. M. Ashworth and J. E. Smith, ed.), pp. 245–267, Cambridge University Press, Cambridge.
BELL, W. and MITCHELL, R. (1972). *Biol. Bull.* **143,** 265–277.
BERG, H. C. and ANDERSON, R. A. (1973). *Nature* **245,** 380–382.
BERG, H. C. and BROWN, D. A. (1972). *Nature* **239,** 500–504.
BERGMAN, K., BURKE, P. V., CERDÁ-OLMEDO, E., DAVID, C. N., DELBRÜCK, M., FOSTER, K. W., GOODELL, E. W., HEISENBERG, M., MEISSNER, G., ZALOKAR, M., DENNISON, D. S. and SHROPSHIRE, W. S. (1969). *Bact. Rev.* **33,** 99–157.
BERGMAN, K., ESLAVA, A. P. and CERDÁ-OLMEDO, E. (1973). *Mol. Gen. Genet.* **23,** 1–16.
BESSIS, M. (1964). *In* "Cellular Injury" (A. V. S. de Reuck and J. Knight, ed.), pp. 287–328. Ciba Foundation Symposium. Churchill, London.
BONNER, J. T. (1952). "Morphogenesis". Princeton University Press, Princeton.
BONNER, J. T. (1967). "The Cellular Slime Molds", 2nd ed. Princeton University Press, Princeton.
BONNER, J. T., CLARKE, W. W., NEELY, C. L. and SLIFKIN, M. K. (1950). *J. Cell. Comp. Physiol.* **36,** 149–158.
BONNER, J. T. and ROSS, M. R. (1962). *Develop. Biol.* **5,** 344–361.
BRETHERTON, F. P. and ROTHSCHILD, LORD (1961). *Proc. Roy. Soc. B,* **153,** 490–502.
BROKAW, C. J. (1958a). *J. Exp. Biol.* **35,** 192–196.
BROKAW, C. J. (1958b). *J. Exp. Biol.* **35,** 197–212.
BROKAW, C. J. (1974). *J. Cell. Physiol.* **83,** 151–158.
CARLILE, M. J. (1965). *Ann. Rev. Plant Physiol.* **16,** 175–202.
CARLILE, M. J. (1966). *In* "The Fungal Spore" (M. F. Madelin, ed.), pp. 175–187. Butterworths, London.
CARLILE, M. J. (1970a). *In* "Photobiology of Micro-organisms" (P. Halldal, ed.), pp. 309–344. Wiley, London.
CARLILE, M. J. (1970b). *J. Gen. Microbiol.* **63,** 221–226.
CARLILE, M. J., DICKENS, J. S. W. and SCHIPPER, M. A. A. (1962). *Trans. Brit. Mycol. Soc.* **45,** 462–464.

CARTER, S. B. (1965). *Nature* **208,** 1183–1187.
CARTER, S. B. (1967). *Nature* **213,** 256–260.
CHET, I., FOGEL, S. and MITCHELL, R. (1971). *J. Bact.* **106,** 863–867.
CHET, I., HENIS, Y. and MITCHELL, R. (1973). *J. Bact.* **115,** 1215–1218.
CHET, I., ZILBERSTEIN, Y. and HENIS, Y. (1973). *Physiol. Plant Path.* **3,** 473–479.
CRICK, F. (1970). *Nature* **225,** 420–422.
CROLL, N. A. (1970). "The Behaviour of Nematodes". Arnold, London.
DELBRÜCK, M. (1972). *Angew. Chem. Internat. Edit.* **11,** 1–6.
DELBRÜCK, M. and REICHARDT, W. (1956). *In* "Cellular Mechanisms in Differentiation and Growth" (D. Rudnick, ed.), pp. 3–44. Princeton University Press, Princeton, New Jersey.
DENNISON, D. S. (1961). *J. Gen. Physiol.* **45,** 23–38.
DENNISON, D. S. and BOZOF, R. P. (1973). *J. Gen. Physiol.* **62,** 157–168.
DENNISON, D. S. and ROTH, C. C. (1967). *Science* **156,** 1386–1388.
DICKINSON, S. (1960). *In* "Plant Pathology—an advance treatise" Vol. 2 (J. G. Horsfall and A. E. Dimond, ed.), pp. 203–232. Academic Press, New York.
DICKINSON, S. (1964a). *Trans. Brit. Mycol. Soc.* **47,** 300–301.
DICKINSON, S. (1964b). Cereal Rust Conferences, June and July 1964.
DOETSCH, R. N. (1972). *J. Theor. Biol.* **35,** 55–66.
DUNN, G. A. (1973). *In* "Locomotion of Tissue Cells" (R. Porter and D. W. Fitzsimmons, ed.), pp. 211–232. Ciba Foundation Symposium 14 (new series). Associated Scientific Publishers, Amsterdam.
DWORKIN, M. (1973). *In* "Microbial Differentiation", Symposium of the Society for General Microbiology, Vol. 23 (J. M. Ashworth and J. E. Smith, ed.), pp. 125–142. Cambridge University Press, Cambridge.
DRYL, S. (1973). *In* "Behaviour of Micro-organisms" (A. Perez-Miravete, ed.), pp. 16–30. Plenum Press, London.
ECKERT, R. (1972). *Science* **176,** 473–481.
EDELSTEIN, B. B. (1971). *J. Theor. Biol.* **30,** 515–532.
FLUEGEL, W. (1963). *Proc. Minn. Acad. Sci.* **30,** 120–123.
FOSTER, K. W. and LIPSON, E. D. (1973). *J. Gen. Physiol.* **62,** 590–617.
FRAENKEL, G. S. and GUNN, D. L. (1961). "The Orientation of Animals". Dover Publications, New York.
FRANCIS, D. W. (1964). *J. Cell Comp. Physiol.* **64,** 131–135.
GOODAY, G. W. (1973). *In* "Microbial Differentiation" (J. M. Ashworth and J. E. Smith, ed.), 23rd Symposium of the Society for General Microbiology, pp. 269–294, Cambridge University Press, Cambridge.
HARRIS, A. K. (1973). *In* "Locomotion of Tissue Cells" (R. Porter and D. W. Fitzsimmons, ed.), pp. 357–361. Ciba Foundation Symposium 17 (new series), Associated Scientific Publishers, Amsterdam.
HAUPT, W. (1965). *Ann. Rev. Plant Physiol.* **16,** 267–290.
HAUPT, W. and SCHÖNBOHM, E. (1970). *In* "Photobiology of Micro-organisms" (P. Halldal, ed.), pp. 283–307. Wiley, London.
INGOLD, C. T. (1953). "Dispersal in Fungi". Clarendon Press, Oxford.
JAFFE, L. (1958). *Exp. Cell Research* **15,** 282–299.
JAFFE, L. and ETZOLD, H. (1962). *J. Cell. Biol.* **13,** 13–31.
JAFFE, L. and ETZOLD, H. (1965). *Biophys. J.* **5,** 715–742.

Jahn, T. L. and Bovee, E. C. (1964). *In* "Biochemistry and Physiology of Protozoa", Vol. 3 (S. H. Hutner, ed.), pp. 61–129. Academic Press, New York.
Jan, Y. N. (1974). *J. Biol. Chem.* **249,** 1973–1979.
Jennings, H. S. (1962). "Behaviour of the Lower Organisms". Indiana University Press, Bloomington.
Jeon, K. W. and Bell, I. G. E. (1965). *Exp. Cell Res.* **38,** 536–555.
Jesaitis, A. J. (1974). *J. Gen. Physiol.* **63,** 1–21.
Johnson, D. L. and Gamov, R. I. (1971). *J. Gen. Physiol.* **57,** 41–49.
Johnson, D. L. and Gamov, R. I. (1972). *Plant Physiol.* **49,** 898–903.
Kamiya, N. (1959). Protoplasmotologia, Vol. 8, pt 3a. Springer, Wien.
Katsura, K. and Miyata, Y. (1971). *In* "Morphological and Biochemical Events in Plant-Parasite Interaction" (S. Akai and S. Ouchi, ed.), pp. 107–128. The Phytopathology Society of Japan, Tokyo.
Kazama, F. Y. (1972). *J. Gen. Microbiol.* **71,** 555–566.
Korohoda, W., Rakoczy, L. and Walczak, T. (1970). *Acta Protozool.* **7,** 363–373.
Koshland, D. E. (1974). *FEBS Letters* **40** (suppl.) 3–9.
Kufferath, H. (1911). *Ann. Inst. Pasteur* **25,** 601–648.
Larsen, S. H., Reader, R. W., Kort, E. N., Tso, W. W. and Adler, J. (1974). *Nature* **249,** 74–77.
Leedale, G. F. (1967). "Euglenoid Flagellates". Prentice Hall, Englewood Cliffs.
Loeb, J. (1973). "Forced Movements, Tropisms and Animal Conduct". Dover, New York.
Loomis, W. F. (1972). *Nature New Biol.* **240,** 6–9.
Machlis, L. (1973a). *Plant Physiol.* **52,** 527–530.
Machlis, L. (1973b). *Plant Physiol.* **52,** 524–526.
Machlis, L. and Rawitscher-Kunkel, E. (1963). *Int. Rev. Cytol.* **15,** 97–138.
Magnus, M. A. and Wolken, J. J. (1974). *Plant Physiol.* **53,** 512–513.
Maheshwari, R. and Hildebrandt, A. C. (1967). *Nature* **214,** 45–46.
Margulis, L. (1970). "Origin of Eukaryotic Cells". Yale University Press, New Haven.
Mascarenhas, J. P. (1973). *In* "Behaviour of Micro-organisms" (A. Perez-Miravete, ed.), pp. 62–69. Plenum Press, London.
Mascarenhas, J. P. and Machlis, L. (1962a). *In* "Vitamins and Hormones" Vol. 20, pp. 347–372. Academic Press, New York.
Mascarenhas, J. P. and Machlis, L. (1962b). *Nature* **196,** 292–293.
Mascarenhas, J. P. and Machlis, L. (1964). *Plant Physiology* **39,** 70–77.
Masson, J. J., Iggo, A., Reid, R. A. and Mann, G. F. (1952). *J. Roy. Microscop. Soc.* **72,** 67–69.
McNab, R. and Koshland, D. E. (1972). *Proc. Nat. Acad. Sci. U.S.A.*, **69,** 2509–2512.
McNab, R. and Koshland, D. E. (1974). *J. Mol. Biol.* **84,** 399–406.
McVittie, A. and Zahler, S. A. (1962). *Nature* **194,** 1299–1300.
Miles, C. A. and Holwill, M. E. J. (1969). *J. Exp. Biol.* **50,** 683–687.
Miller, R. L. (1966). *J. Exp. Biol.* **162,** 23–44.

MILLER, R. L. (1973). *In* "Behaviour of Micro-organisms" (A. Perez-Miravete, ed.), pp. 31–47. Plenum Press, London.
MILLER, R. L. and BROKAW, C. J. (1970). *J. Exp. Biol.* **52,** 699–706.
MITCHELL, R., FOGEL, S. and CHET, I. (1972). *Water Research* **6,** 1137–1140.
MULLER, D. and JAFFE, L. (1965). *Biophys. J.* **5,** 317–355.
NAITOH, Y. and ECKERT, R. (1974). *In* "Cilia and Flagella" (M. A. Sleigh, ed.), pp. 305–352. Academic Press, London.
NAITOH, Y. and KANEKO, H. (1972). *Science* **176,** 523–524.
ORTEGA, J. K. E. and GAMOV, R. I. (1970). *Science* **168,** 1374.
PETERSEN, J. E. (1969). *In* "Methods on Microbiology", Vol. 3B (J. R. Norris and D. W. Ribbons, ed.), pp. 185–210. Academic Press, London.
PETZUCH, M. and DELBRÜCK, M. (1970). *J. Gen. Physiol.* **56,** 297–308.
POFF, K. L., BUTLER, W. L. and LOOMIS, W. F. (1973). *Proc. nat. Acad. Sci. U.S.A.* **70,** 813–816.
RAKOCZCY, L. (1963). *Acta Soc. Bot. Pol.* **32,** 393–403.
RAMSEY, W. S. (1972). *Exp. Cell Res.* **70,** 129–139.
ROBERTSON, J. A. (1972). *Arch. Mikrobiol.* **85,** 259–266.
ROBINSON, P. M. (1973). *New Phytol.* **72,** 1394–1356.
ROSEN, W. G. (1962). *Quart. Rev. Biol.* **37,** 242–259.
ROSEN, W. G. (1968). *Ann. Rev. Plant Physiol.* **19,** 435–462.
ROSENBERG, B., MISRA, T. N. and SWITZER, R. (1968). *Nature* **217,** 423–427.
ROTHSCHILD, LORD (1956). "Fertilization" Methuen, London.
ROYLE, D. J. and HICKMAN, C. J. (1964). *Can. J. Microbiol.* **10,** 151–162.
RUHLAND, W. ed. (1959, 1962). "Encyclopedia of Plant Physiology", Vol. 17, parts 1 and 2. Springer: Berlin.
SCHAFFER, B. M. (1964). *In* "Primitive Motile Systems in Cell Biology (R. D. Allen and N. Kamiya, ed.), pp. 387–405. Academic Press, New York.
SCHEIDER, O. (1968). *Z. Pflanzenphysiol.* **59,** 258–273.
SEYMOUR, F. W. K. and DOETSCH, R. N. (1973). *J. Gen. Microbiol.* **78,** 287–296.
SILVERMAN, M. and SIMON, M. (1974). *Nature* **249,** 73–74.
SMITH, J. L. and DOETSCH, R. N. (1969). *J. Gen. Microbiol.* **55,** 379–391.
STADLER, D. R. (1952). *J. Cell. Comp. Physiol.* **39,** 449–474.
STANIER, R. Y. (1942). *J. Bact.* **44,** 405–412.
STANIER, R. Y. (1947). *Nature* **159,** 682–683.
STANIER, R. Y. (1974). *In* "Evolution in the Microbial World" (M. J. Carlile and J. J. Skehel, ed.). 24th Symposium of the Society for General Microbiology, pp. 219–240. Cambridge University Press, Cambridge.
TROUTMAN, J. L. and WILLS, W. H. (1964). *Phytopathology* **54,** 225–228.
TSANG, N., MCNAB, R. and KOSHLAND, D. E. (1973). *Science* **181,** 60–63.
TSO, W. W. and ADLER, J. (1974). *J. Bact.* **118,** 560–576.
WAALAND, S. D. and CLELAND, R. E. (1974). *Protoplasma* **79,** 185–196.
WARD, S. (1973). *Proc. Nat. Acad. Sci. U.S.A.* **70,** 817–821.
WEISS, P. (1945). *J. exp. Zool.* **100,** 353–386.
YOUNG, L. Y. and MITCHELL, R. (1973). *Applied Microbiol.* **25,** 972–975.
ZURZYCKI, J. (1972). *Acta Protozool.* **11,** 189–199.

Chapter 2

Phototaxis and photokinesis

W. NULTSCH

Department of Botany, University of Marburg, Germany

I. Introduction

Phototactic reactions in micro-organisms were first detected by Treviranus (1817). He observed that swarmspores of several species of algae accumulated either at the illuminated side of a vessel or at the opposite side, as a result of a migration either towards the light source or away from it. A different type of reaction was described by Engelmann (1882). He found that in the so-called *Bacterium photometricum*, probably a species of the purple sulphur bacterium genus *Chromatium*, an abrupt backward movement was caused by a sudden decrease in light intensity. Since the organisms behaved as if they were frightened, he called this reaction a phobic response ("Schreckbewegung"). An effect of light on motility had already been reported by Strasburger (1878), who observed a cessation of movement at high light intensities in green swarmspores. In Engelmann's (1882, 1883) experiments the purple bacteria became motionless in the dark but resumed their movement when the light was switched on again. Engelmann called this phenomenon photokinesis. These three reaction types were later found to occur also in many other motile micro-organisms—see the reviews of Bendix (1960), Checcucci (1973), Clayton (1959, 1964), Davenport (1973), Feinleib and Curry (1971a), Halldal (1962, 1964), Hand and Davenport (1970), Haupt (1959, 1965, 1966), Nultsch (1970, 1973a, b, 1974a, b) and Tollin (1969).

The term "photokinesis" has been retained but the terminology of the other reaction types has changed repeatedly. Rothert (1901) distinguished between "strophic" (photo-orientation) and "apobatic" (phobic) phototaxis. Nagel (1901) restricted the term "phototaxis" to the *orientation* of movement by light and introduced the term "discrimination sensitivity" instead of "phobic" response. Massart (1902) proposed "phobism" instead of "Schreckbewegung". Pfeffer (1904) suggested the terms "photo-topotaxis" for the photo-orientation of movement and "photo-phobotaxis" for the "phobic response". These last two terms have been widely accepted and have been used by most authors in the following decades. Usually the words "positive" and "negative" are added to indicate the sense of the reaction.

Recently, however, new attempts have been made to change the terminology. Diehn (1970, 1973) took up Nagel's suggestion of restricting the term "taxis" to orientated movement and re-introduced Engelmann's "phobic response", distinguishing between direct and inverse responses instead of negative and positive ones, the direct response being caused by an increase and the inverse response by a decrease in light intensity. Hand and Davenport (1970), who also restricted "taxis" to the directed movements, suggested the terms "klinokinesis" and "orthokinesis" for "phobotaxis" and "kinesis" respectively. Finally, Uffen *et al.* (1971) suggested the term "skotophobic" response for positive photophobotactic response, so that "photophobic" would be restricted to negative photo-phobotactic reaction, which seems to be more logical than the terms inverse and direct photophobic response.

These disagreements in terminology have, in the past, led to errors and misinterpretation, so that the creation of new terms would further confuse the situation rather than clarify it. Furthermore, whether the term "photo-phobotaxis" is a more "uncomfortable mouthful" (Hand and Davenport, 1970) than "photo-klinokinesis" is disputable. The Greek word "taxis" in its original sense denoted only a distinct spatial array, and so the term "phototaxis" should mean any array of organisms in space caused by light and is not restricted to their directed movement. Therefore, Pfeffer was correct to add the prefixes "topo" and "phobo", since they give further information as to how this array is brought about, although arrays due to photokinetic effects were not considered. Hence the term photo-phobotaxis is not self-contradictory as was asserted by Checcucci (1973) and Diehn (1973).

Since none of the suggested terms are wholly adequate and universally accepted, the author of this chapter exercises some freedom in the use of terms, although in general Pfeffer's terminology is followed. In spite of the confusion in terminology, the existence of three different photic reaction types is clear. They can be defined as follows:

1. *Photo-topotaxis* denotes a type of photomotion in which the direction of movement depends on the direction of the incident light beam, and results in a migration either towards the light source (positive reaction) or away from it (negative reaction). It can be brought about either by a true steering act, i.e. by a deviation from the original course and an active orientation to the light direction, or by changing the rhythm of non-orientated backward and forward movement, in which the orientation with respect to light direction is random, but movement in one direction, either towards the light source or away from it, is preferred. An indispensible prerequisite for phototactic orientation is the ability of the organism to determine its position in relation to the light direction by measuring and comparing differences in the fluxes of quanta which result from the absorption gradient under unilateral illumination. This can be brought about in two different ways: either by comparison of simultaneous measurements with the aid of two or more photoreceptors, or by modulation. The latter term refers to the comparison of successive measurements made by a single photoreceptor, the position of which in relation to the light source is changed, e.g. by rotation. If the difference measured exceeds a certain value, a motor response is triggered to correct the course. Thus phototactic orientation requires the repeated comparison of at least two either simultaneous or successive measurements of light intensity.

2. *Photo-phobotaxis*. Photo-phobotactic reactions, also called "phobic responses", "shock reactions", "stop responses", "motor responses" or "skotophobic responses" are caused by temporal changes in irradiance, dI/dt, independent of the direction of the incident light, which can impinge upon the organism from two opposite sides or be diffuse. Thus, photo-phobotactic reactions differ from photo-topotactic reactions in

being single transient motor responses, each caused by a sudden change in the flux of quanta, either a decrease (positive reactions) or an increase (negative reactions). When a spatial gradient of illuminance is established, any organism traversing this gradient perceives a temporal change in intensity, since with movement the spatial gradient is transformed into a temporal one. Thus, photophobic responses can be induced at defined places, eventually resulting in accumulations of organisms along borders between areas of different light intensities, in light fields (positive reactions) or in migrations out of the light field in case of negative reactions. These various consequences of the photophobic responses are all classified as photo-phobotaxis.

3. *Photokinesis* means an effect of light on motility, consisting either of an increase (positive) or decrease (negative) in linear velocity. Under certain conditions, photokinetic effects can result also in the starting or stopping of movement.

II. Photo-topotaxis

A. *Methods of study*

Various methods have been used to quantify photo-topotactic reactions. These range from simple microscopic observations to modern electronic techniques. According to Hand and Davenport (1970), two main groups can be distinguished:

(1) "mass movement" or population methods in which the behaviour of whole cell populations is investigated, and
(2) "individual cell" methods in which the behaviour of single cells is studied.

In addition, a third approach can be distinguished, a cell counting method in which conclusions are drawn from records on many individuals.

1. *Population methods*

Two essentially different approaches are used to determine the effect of unilateral illumination: the measurement of differences in optical density due to changes in the distribution of organisms in an originally homogeneous suspension (methods a–e below) and the analysis of the migration of cell masses (methods f and g).

(a) *Rotating cuvette*. Feinleib and Curry (1967) used a cuvette which was rotated continuously about its longitudinal axis at 60 rpm. As shown in Fig. 1, the cuvette is placed in a horizontal position so that the actinic (stimulating) light beam enters it from the front end which is closed by a cover glass. The rear end is closed by black paraffin to diminish reflection. The measuring beam of red light ($\lambda > 620$ nm), forming a right angle to the actinic beam, passes through the cuvette containing the algal suspension and a second red filter to remove scattered actinic light, and impinges on two photocells connected in a comparison circuit. The voltage difference between the photocells, which is zero at the beginning of the experiment, is measured and plotted continuously. The magnitude of the signal is directly proportional to the number of organisms and serves as a measure of the phototactic response.

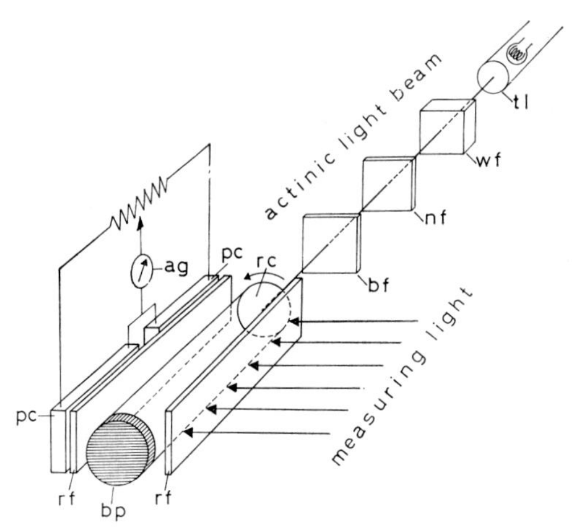

FIG. 1. The apparatus used by Feinleib and Curry (1968) to measure photo-topotactic reactions. ag = amplifier-galvanometer, bf = blue filter, bp = black paraffin, nf = neutral density filter, pc = photocells, rc = rotating cuvette, rf = red filter, tl = tungsten lamp, wf = water filter.

This method, also used by Marbach and Mayer (1970), is well suited for studying the influence of external and internal factors on the phototactic response. However, it has some disadvantages:

(1) Inside the cuvette a light gradient exists due to the absorption and scattering of the actinic light by the cells. Thus it can happen that the cells react negatively at the illuminated end but positively in the rear, resulting in accumulations in the middle part of the cuvette.
(2) In photosynthetic organisms positive phototaxis can be simulated by positive chemotaxis to oxygen, the concentration of which at the illuminated end increases during an experiment due to the higher rate of photosynthesis in this part of the cuvette. Simultaneously, an inverse CO_2 gradient originates which may also influence the reaction.
(3) If the same preparation is used for a long time the conditions inside the cuvette change considerably and may influence the results of the experiments.

(b) 45° *illumination.* The first two of the above disadvantages are avoided when a non-rotating cuvette with photocells at both ends is arranged at an angle of 45° to the actinic light beam (Fig. 2) as suggested by Throm (1968). If parallel light is used, the illuminance is equal throughout the length of the cuvette, and selfshading is negligible when the diameter of the cuvette is small. The formation of chemical gradients by photosynthesis is therefore impossible. The organisms move at 45° either towards the light source or away from it and accumulate at one end of the cuvette, causing a difference in output between the two photocells mounted near the ends of the cuvette. The cuvette can be used as a flow

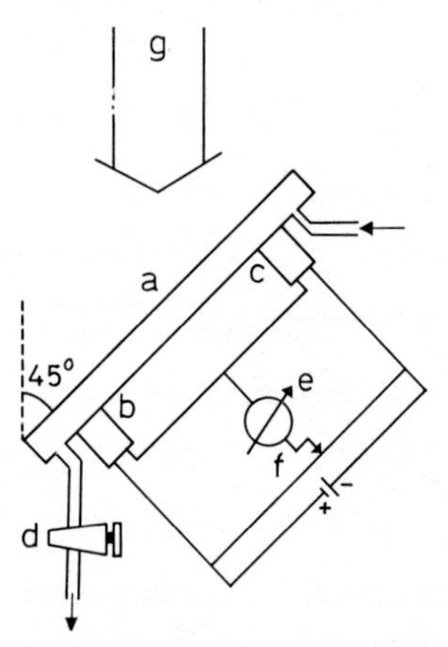

FIG. 2. The 45° illumination technique. a = flow through cuvette, b, c = photo resistors d = stop cock, e = Kipp micrograph BD, f = variable resistance, g = actinic light. After Throm (1968).

through system. The main disadvantage of the method is that the actinic light serves simultaneously as the measuring light.

(c) *Automatic phototaxis monitoring device.* All the above sources of error are eliminated in an automatic phototaxis monitoring device (Nultsch, Throm and v. Rimscha, 1971; Nultsch and Throm, 1975). In this apparatus (Fig. 3) the cuvette is placed horizontally so that its long axis forms an angle of 45° with the actinic light beam while the measuring light (> 585 nm) comes from above. Under the cuvette are mounted two photoresistors connected in a bridge circuit. On a given impulse, the cell suspension is pumped from a continuous culture into a collecting bottle, from which a steady stream of suspension, freed from air bubbles, flows through the cuvette back into the culture vessel. Since the suspension is homogeneous and the organisms are mechanically prevented by the stream from reacting phototactically, the two photoresistors are illuminated equally and the recorder yields an almost straight base line.

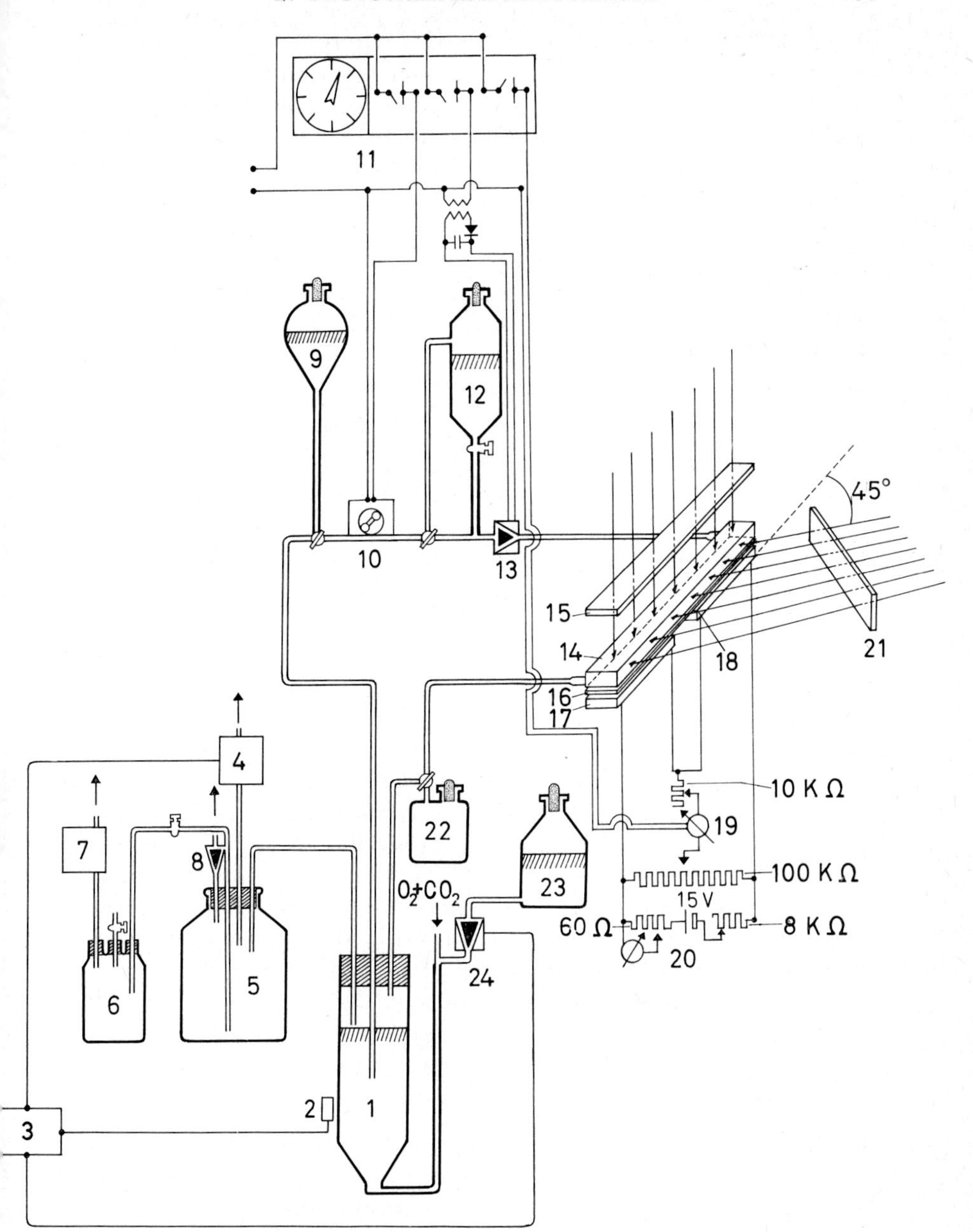

FIG. 3. Improved automatic phototaxis monitoring device. 1. Culture vessel, 2. photocell, 3. dilution control device, 4. pump, 5, 6. collecting bottles, 7. pump, 8. valve, 9. rinsing water vessel, 10. pump, 11. timer, 12. collecting vessel, 13. magnetic valve, 14. swimming cuvette with inlet and outlet tube, 15. measuring light red filter, 16. red filter, 17, 18. photoresistors, 19. Kipp-micrograph, 20. bridge circuit, 21. actinic light filter, 22. rinsing water collecting vessel, 23. nutrient medium storage bottle, 24. magnetic valve. After Nultsch and Throm (1975).

When the flow of suspension is stopped by closing a magnetic valve, the organisms migrate towards the light source, causing an asymmetry in the optical density. A Kipp micrograph records the time course of the phototactic response. After the experiment is finished, the circulation system is switched on automatically, removing the inhomogeneous suspension from the cuvette. An experiment can be run every 15 minutes. The only disadvantage of this system is that large organisms such as *Euglena* sometimes settle out. Pattern formation in the cell concentrations used (Nultsch and Hoff, 1973) is suppressed by phototactic movement. If the measuring light, which is photokinetically active, is strong enough, the photokinetic reaction, which could change the slope of the recordings, is saturated and does not invalidate the results.

(d) *Light trap methods*. The light trap method, although devised to study photo-phobotaxis (Engelmann, 1882; Buder, 1915), has been used to measure photo-topotactic reactions too. Bünning and Schneiderhöhn (1956) observed in light trap experiments that *Euglena* cells in the vicinity of the light field were attracted topotactically by laterally scattered light. They used the number of cells entering the light field per unit time as a measure of photo-topotactic activity. Lindes *et al.* (1965) measured the optical density of cell accumulations in a light trap with the so-called phototaxigraph (see page 52) and used it to quantify photo-topotaxis. Stavis and Hirschberg (1973) have used a similar apparatus.

Since accumulations in a light trap are brought about mainly by photo-phobotaxis and, under certain conditions, by negative photokinesis (Clayton, 1957, 1964), it may happen that an overall photomotion response is measured this way. Thus, misinterpretations cannot be excluded (see page 52). Furthermore, the actual light source for topotaxis in this system is the organisms in the light trap from which light is scattered. Since their number changes during an experiment, we know neither the intensity of the scattered light nor, when white light or broad bands of coloured light are used, its spectral composition. Therefore, this technique should not be used for investigations of photo-topotactic responses.

(e) *Projected spectrum technique*. Halldal (1958) devised an apparatus for measuring photo-topotactic action spectra (Fig. 4), in which a spectrum is projected onto a rectangular cuvette, containing the cell suspension. A wedge is inserted into the beam to produce an intensity gradient at a right angle to the wavelength scale. From the opposite side, the cuvette is irradiated with a reference beam of constant wavelength and intensity. If the effectiveness of the actinic light is low, the organisms will be more attracted by the reference light and will accumulate at the wall nearest the reference light source. When the actinic light is more effective, the cells will accumulate at the wall facing this source. When the reference and actinic beams are of approximately equal effectiveness the organisms will display random motion. If the reference beam is properly chosen in wavelength and intensity and if the spectrum is adjusted to equal numbers of quanta at each wavelength, the region of random motion forms a curve representing the action spectrum of photo-topotaxis. This curve will constitute the upper limit of the accumulation of organisms on the cuvette wall facing the actinic beam. The organisms at the wall can be made visible with a narrow white measuring light beam (indicated by the dashed lines with arrows in Fig. 4) after the other light sources have been switched off. This technique enables the investigator to measure an action spectrum in one exposure. Clayton (1964, p. 55, p. 67) have questioned whether the pattern of accumulation is brought

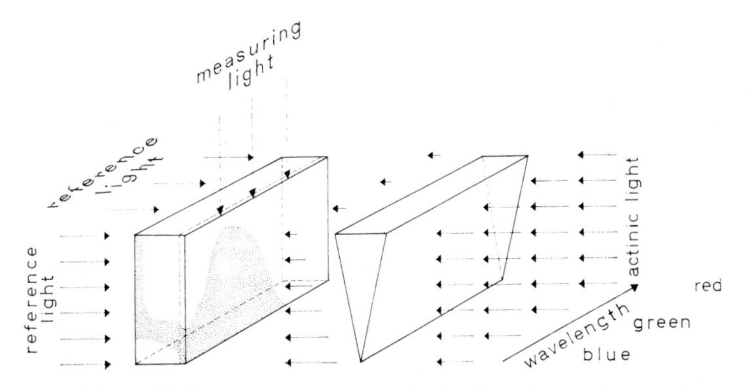

FIG. 4. Halldal's (1958) projected spectrum technique. For explanation see text.

about solely by photo-topotaxis. It must be emphasized that in Halldal's apparatus the cells do not gather in a light gradient, but discriminate between two opposite light beams, the effectiveness of one depending on both wavelength and intensity.

(f) *Measurement in gliding cell masses.* In slowly gliding organisms which move directly towards the light source, such as blue-green algae of the genus *Anabaena* (Drews, 1959), some Chroococcales (Stanier *et al.*, 1971) and the red alga *Porphyridium* (Pringsheim, 1968), the change in direction of movement under unilateral illumination can be used to measure photo-topotactic activity. A similar approach was used by Francis (1964) with *Dictyostelium.*

Moreover the distance traversed by the organisms in a given time, either towards the light source or away from it, can be measured. In this case it must be taken into account that the distance depends also on the speed of movement, which in turn can be influenced by the stimulus light when it is photokinetically active. Therefore Nultsch (unpublished), measuring the photo-topotactic action spectrum of *Porphyridium cruentum*, saturated photokinesis by using a photokinetically active but phototactically ineffective background light impinging from above.

(g) *Measurement of spreading areas*. In some other slowly moving organisms, such as diatoms (Nultsch, 1971) and blue-green algae of the genus *Phormidium* (Drews, 1959; Nultsch, 1961), the spreading area is shifted towards the light source in case of positive reactions (Fig. 5) and away

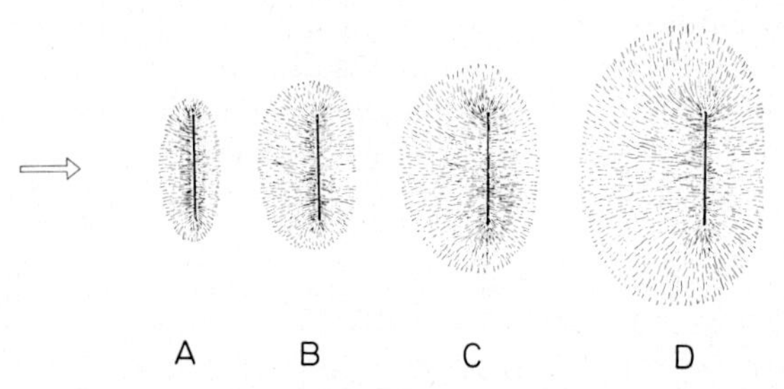

Fig. 5. Asymmetric spreading of *Phormidium uncinatum* about 1, 2, 3 and 5 hours (A-D) after beginning unilateral illumination (arrow). After Drews (1959).

for negative reactions. The relative extent of shifting can be used as a measure of the photo-topotactic effect (Nultsch, 1961, 1971).

2. *Individual cell methods*

The main advantage of the population methods is that they depend on the average behaviour of a great number of cells and aberrant behaviour by individuals does not influence results. However, they give no information on how the orientation of the cells is brought about. Behavioural studies of individual cells are therefore important in phototactic investigations.

(a) *Direct microscopic observations*. This is the most obvious way of studying the behaviour of individual cells. For slowly moving organisms time-lapse and for swift ones high speed cinematography is required.

(b) *Flying spot scanning device*. To monitor the movement of several cells simultaneously, Davenport *et al.* (1962) and Hand *et al.* (1965) designed the "flying spot scanning device". In this apparatus (Fig. 6) a light beam generated by an oscilloscope scans a sample. The absorption changes due to the moving organisms are detected by a photomultiplier and transformed into electrical potentials, which are amplified and transmitted to a second oscilloscope. This way the tracks of the moving organisms can be followed visually or recorded photographically. A typical track recorded this way is shown in Fig. 7. The method is applicable to organisms of diameter greater than 15–20 μm. A disadvan-

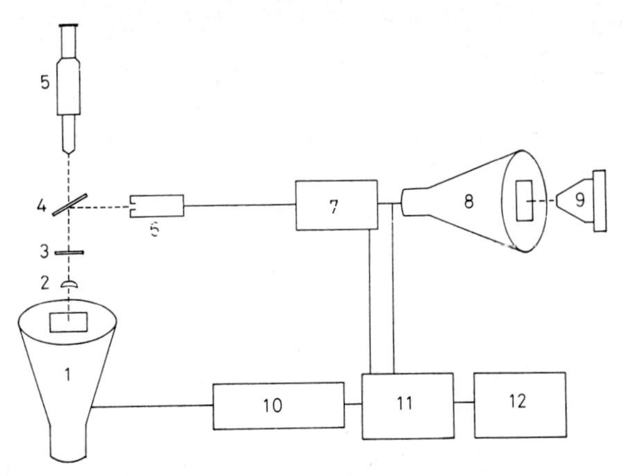

FIG. 6. The flying spot scanning device. 1. Scanning tube, 2. lens, 3. sample, 4. half mirror, 5. microscope, 6. photomultiplier, 7. video amplifier, 8. display tube, 9. camera, 10. adjustable time interval gate, 11. time base, 12. regulated power supply. With permission from Hand and Davenport (1970).

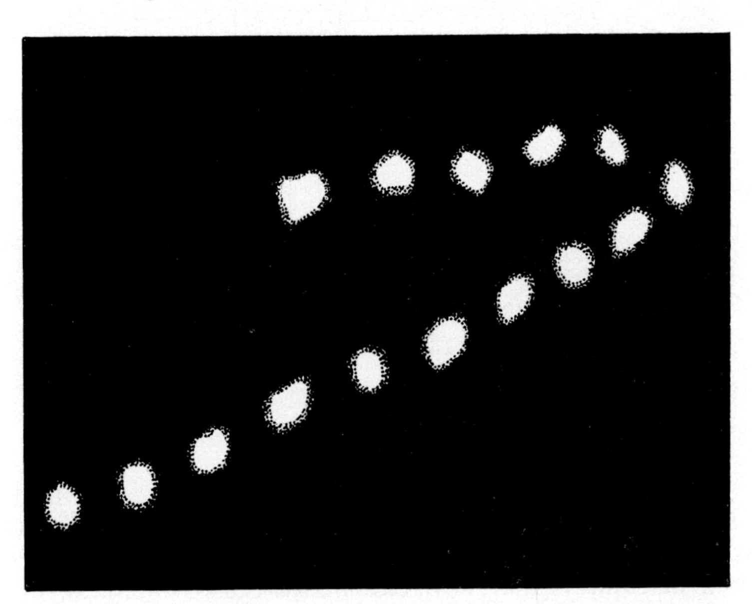

FIG. 7. Typical track of the dinoflagellate *Gyrodinium* as photographed through the flying spot scanning device. Modified from Hand and Davenport (1970).

tage is that the scanning wavelength which cannot be varied might be such as to influence behavior. In an improved system Davenport *et al.* (1970) used a television camera sensitive to all visible wavelengths to view the optical image directly through the cine attachment of the microscope, and could adjust illuminating wavelengths with filters.

The movement of organisms can be observed either directly on the viewing screen or the images can be stored on a video tape. To reduce the inflow of data to the computer to those pictures which are of interest, a "bug-watcher" (Fig. 8) was used between the video information source and the computer, selecting the pictures according to their level of grayness. The traces obtained this way resemble the one in Fig. 7. For more details see Davenport (1973).

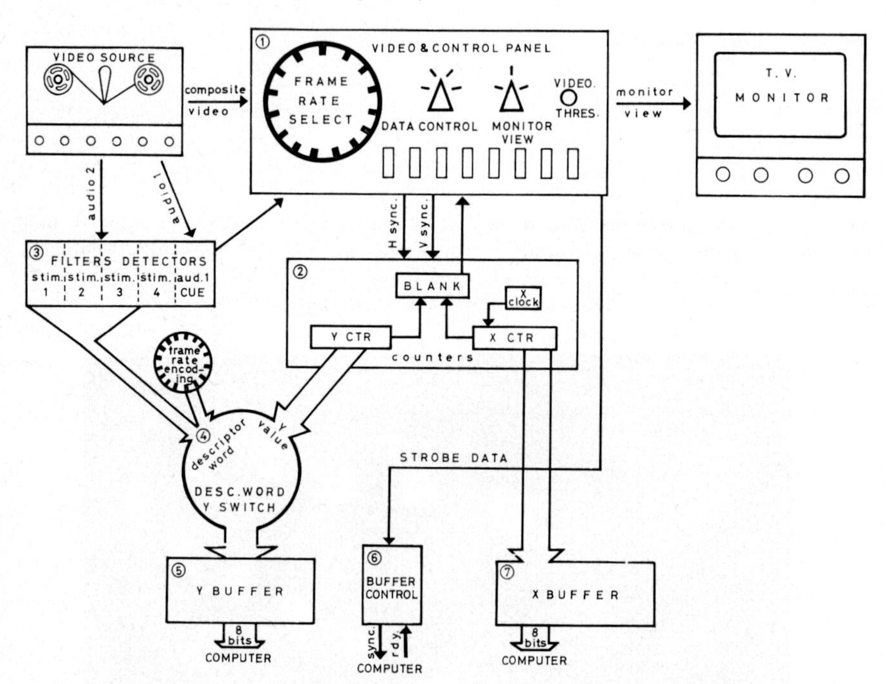

FIG. 8. Block diagram of the "bugwatcher". With permission of Davenport *et al.* (1970).

(c) *Analysis of tracks.* Nultsch (1956) projected the microscope field onto a table and traced the tracks of diatoms on paper, and Neuscheler (1967a) dyed the mucilage traces made by *Micrasterias* cells (Fig. 9). Feinleib and Curry (1967) with *Chlamydomonas* used the tracks recorded on a film during a fixed exposure time (see page 70f.)

3. *Cell counting methods*

Cell counting techniques can be used to determine the number of cells swimming either towards the light source or away from it. In the technique of Rikmenspoel and van Herpen (1957), which was devised to study the behaviour of bull spermatozoa, images of swimming cells are produced with darkfield illumination in the image plane, where a diaphragm with an aperture of approximately 100 μm is placed. The response of a photomultiplier, mounted so as to receive a light flash whenever a cell passes the aperture, is recorded. The number of cells passing the aperture in a given time under unidirectional illumination could therefore serve as a measure of photo-topotactic activity.

B. Intensity dependence

Most of the organisms investigated so far respond positively in dim light (cf. Haupt, 1959; Nultsch, 1970) provided that a threshold value, the so-called absolute or zero threshold, is exceeded. The following

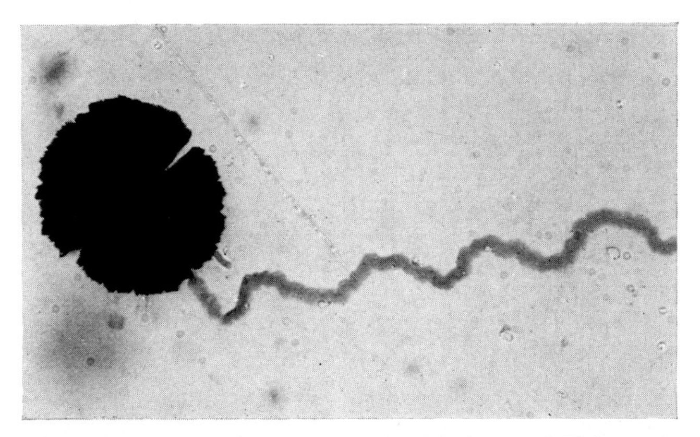

FIG. 9. Dyed mucilage traces of *Micrasterias*. The black area (left) is the dyed *Micrasterias* cell. Reproduced with permission from Neuscheler (1967a).

threshold values were measured: in diatoms 1–30 lux (Heidingsfeld, 1943; Nultsch, 1956, 1971), in blue-green algae 1–30 lux (Drews, 1959; Nultsch, 1962c), in *Chlamydomonas* 1 erg cm^{-2} sec^{-1} (Feinleib and Curry, 1971b), in *Micrasterias* 10^{-4}–10^{-5} lux (Neuscheler, 1967b).

With increasing light intensities the photo-topotactic reactivity, represented either by the number of positively responding cells or by the directness of the swimming paths or both, becomes more pronounced. In this range of the dose-response-curves the relationship between stimulus intensity and response is commonly logarithmic, as shown by Feinleib (see Feinleib and Curry, 1971a) in *Chlamydomonas* and by Diehn and Tollin (1966) in *Euglena*. Phototactic responses adhere to the Weber-Fechner Law over a certain range of intensities.

The optimum (Fig. 10) is reached at different light intensities in different species, e.g. 200 lux in *Nitzschia communis* (Nultsch, 1971), 50–200 lux in several *Phormidium* species (Nultsch, 1962c) and 1,000 lux in *Chlamydomonas reinhardtii* (Nultsch *et al.*, 1971). Above the optimum the positive phototactic effect is reduced by further increase in light intensity, so that the zero response line is crossed, and finally the response becomes negative. At the inversion intensity, also called "Indifferenz-zone" (indifference zone, Buder, 1917), "Umschlagspunkt" (change point, Mainx, 1929) or "Neutralpunkt" (neutral point, Luntz, 1932), the organisms display, according to older reports, random orientation. Feinleib and Curry (1971b), however, observed in *Chlamydomonas* that an individual cell makes an abrupt switch from a positive to a negative response, although the net response of the population may go through an "indifferent" stage.

In various organisms the following inversion intensities were measured: blue-green algae, 1,000–10,000 lux (Nultsch, 1961, 1962c); *Rhodospirillum*

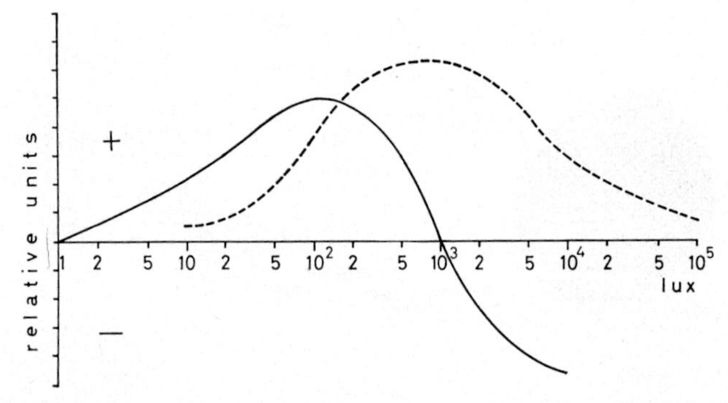

FIG. 10. Photo-topotactic dose-response-curves (white light) of *Phormidium ambiguum* (solid line) and *Chlamydomonas reinhardtii* (dashed line). Abscissa: light intensity in lux; ordinate: phototactic effect in relative units. After Nultsch (1962c) and Nultsch *et al.* (1971).

rubrum, 18,000 lux (Throm, 1968); *Micrasterias*, 4,000 lux (Neuscheler, 1967b). In some organisms, e.g. some diatoms (Nultsch, 1956), negative reactions have never been observed, even at very high intensities. In *Chlamydomonas reinhardtii*, Nultsch *et al.* (1971) also failed to demonstrate negative topotaxis (Fig. 10), although Feinleib and Curry (1971b) found negative responses beyond 10^4 erg cm^{-2} sec^{-1}. This discrepancy may be due to differing conditions in the material used; it is known that threshold values and inversion intensities are influenced by a range of internal and external factors. The complexity of the problem is indicated by the work of Feinleib and Curry (1967) who found that some *Chlamydomonas* cultures responded negatively at higher and positively at lower intensities to the mercury yellow line (577 nm) while others did not.

C. *Action spectra*

In order to find out the chemical nature of the photoreceptor, action spectra were obtained for various organisms. Although some inconsistencies exist, most of the action spectra have one feature in common: only visible light of the shorter wavelengths (up to 550 nm) and in some species near u.v., is active. Maximum activity in flagellated forms is often found between 470 and 500 nm. Similarly, the action spectrum of *Porphyridium cruentum* shows maximum activity between 480 and 540 nm (Nultsch, unpublished). In the diatom *Nitzschia communis* light of shorter wavelengths and especially the near u.v. is more active (Fig. 11). The spectral sensitivity of the colourless zoospores of *Allomyces* (Robertson, 1972) resembles that of *Chlamydomonas* (Fig. 11). The effectiveness of the

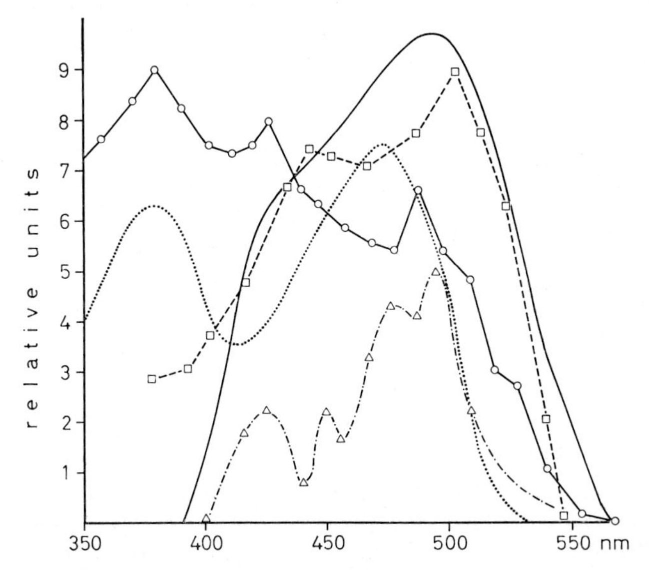

FIG. 11. Photo-topotactic action spectra of *Euglena gracilis* (triangles and dashed-dotted line, Bünning and Schneiderhöhn, 1956), *Euglena gracilis* (dotted line, Diehn, 1969), *Platymonas* (solid line, Halldal, 1958), *Chlamydomonas reinhardtii* (squares and dashed line, Nultsch *et al.*, 1971), *Nitzschia cummunis* (circles and solid line, Nultsch, 1971). Abscissa: wavelength in nm; ordinate: photo-topotactic effect in relative units.

shorter wavelengths has led to the yellow pigments, carotenoids and flavins being considered as possible photoreceptors. The maximum activity of wavelengths around 500 nm in most of the flagellates favours the view that carotenoids or carotenoproteins are the photoreceptor pigments (Bünning and Schneiderhöhn, 1956; Halldal, 1961, 1963; Forward, 1973). However, in the action spectra of *Nitzschia communis* (Fig. 11), *Euglena gracilis* (Fig. 11) and *Phormidium autumnale* (Fig. 12), strong activity is found in the near u.v. between 350 and 400 nm. Since the maxima between 370 and 380 nm coincide with the absorption maximum of riboflavin, Diehn (1969a) has suggested flavins or flavoproteins to function as photoreceptor pigments. As will be shown later, it is possible that both pigments have a role in photo-topotaxis: one as a photoreceptor and the other as a shading pigment.

However, there are also some organisms the action spectra of which do not fit the carotenoid and/or flavin concept well, e.g. the action spectrum of *Prorocentrum micans* shows a maximum between 550 and 600 nm (Fig. 12). Even if we follow Krinsky and Goldsmith (1960), who explained the shift to longer wavelengths by suggesting a hydroxy-echinenone as photoreceptor, or Edmondson and Tollin (1971), who have found that some flavoprotein complexes absorb also in this spectral region, the complete inactivity in the range 400–500 nm cannot be

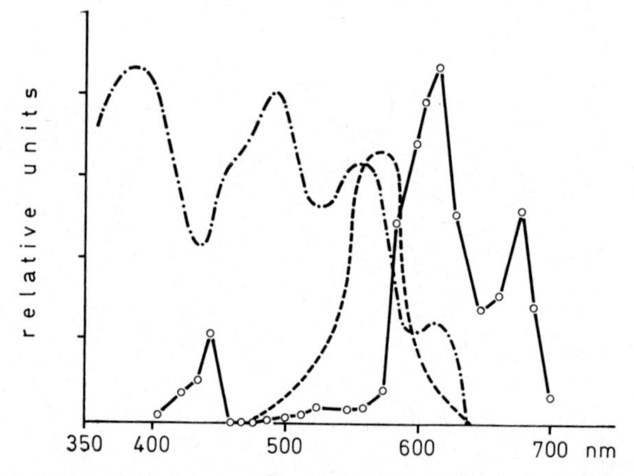

Fig. 12. Photo-topotactic action spectra of *Phormidium autumnale* (dashed-dotted line), *Prorocentrum micans* (dashed line) and *Anabaena variabilis* (circles and solid line). Abscissa: wavelength in nm; ordinate: phototactic effect in relative units. After Nultsch (1961), Halldal (1958) and Nultsch (unpublished).

explained in this way. The action spectrum of the colourless pseudo-plasmodia of *Dictyostelium discoideum* (Francis, 1964; Poff *et al.*, 1973) also cannot easily be interpreted by assuming carotenoids or flavins as the sole photoreceptor pigments. Perhaps the phototactic response of *Dictyostelium* is controlled by a composite system consisting of a photo-receptor pigment, possibly a flavin, which maximally absorbs at about 465 nm and an accessory photosensitive pigment, perhaps a *b*-type cytochrome, which is responsible for the absorption changes detected at 428 and 445 nm (Poff and Butler, 1974). In the blue-green algae *Phormidium uncinatum* and *Phormidium autumnale* the action spectra (Fig. 12) extend to 640 nm (Nultsch, 1961, 1962c). Although the strong effectiveness of radiation between 450 and 510 nm points to yellow pig-ments and the peak in the u.v. favours flavins being the photoreceptors, the maximum around 560 nm and the shoulder at 615 nm indicate that the biliproteins C-phycoerythrin and C-phycocyanin are active. However, correlations between photo-topotaxis and photosynthesis can be excluded in these organisms because red light absorbed by chlorophyll *a* is quite ineffective. This seems not to be true in all blue-green algae. Already Drews (1959) has observed that *Oscillatoria mougeotii* reacts topotactically even in red light up to about 710–730 nm. Most recently, Nultsch (unpublished) has measured the topotactic action spectrum of *Anabaena variabilis*. As shown in Fig. 12, he found maximum activity at 613 nm indicating C-phycocyanin to be the main photoreceptor pigment and smaller but distinct maxima around 430 and 670 nm, pointing to chlorophyll *a*. To our knowledge this is the first action spectrum that indicates correlations between photo-topotaxis and photosynthesis. Of course, more detailed investigations are necessary in this field.

All the above mentioned action spectra were measured in positively reacting organisms. Halldal (1961) compared the action spectra of positive and negative photo-topotaxis in *Platymonas subcordiformis*. Since they turned out to be essentially identical, it seems to be proved that at least in this organism positive and negative phototactic responses are mediated by the same photoreceptor pigment.

D. *Mechanisms*

Another problem is how phototactic orientation is brought about under unilateral illumination, or, in modern jargon, how the stimulus is transduced from the photoreceptor to the effector. It is improbable that only one type of sensory transduction has developed for photo-topotaxis in micro-organisms, since at least two different reaction mechanisms exist:

1. Active steering.
2. A change in the autonomous rhythm of reversal on exposure to unilateral light.

1. *Steering mechanisms*

As mentioned above, the term "steering mechanism" implies an active orientation to the direction of the light beam. However, this active orientation can be brought about in different ways.

(a) *Euglena*. In 1906, Jennings suggested that the phototactic orientation of *Euglena* is the result of numerous, successive "phobic" responses, each of them caused by a slight sideward beating of the flagellum due to a transient shading of the photoreceptor. Mast (1911) and Buder (1917) observed that the movement direction was changed just when the cell rotating around its longitudinal axis brought the stigma (an area of pigment) to the side nearest to the light source. This so-called periodical shading theory (it was postulated that the stigma shaded the photoreceptor) although modified in some respects, has been widely accepted. Diehn and Tollin (1966) and Diehn (1969c) have reported some experimental results supporting the idea. The theory assumes the existence of two pigment systems, consisting of a photoreceptor near the flagellar base and a shading pigment in the stigma (Vavra, 1956; but see the discussion in the review of Feinleib and Curry, 1971a). According to Diehn (1969a) and Tollin (1969) the photoreceptor is a flavoprotein probably embedded in a lipid matrix (Diehn and Kint, 1970; Froehlich and Diehn, 1974) which is located in the paraflagellar swelling, while the stigma, functioning as shading device, contains large amounts of carotenoids, mainly lutein and cryptoxanthin (Batra and Tollin, 1964).

(b) *Other flagellates.* The assumption that the stigma functions as a shading device of the photoreceptor might be right or wrong in case of *Euglena*, but it cannot explain the mechanism of phototactic orientation in flagellates which either lack a stigma, or in which periodical shading of the flagellar base by the stigma can be excluded for morphological reasons. Metzner (1929) observed that the phototactic orientation in *Peridinium* and *Ceratium* is also the result of numerous successive steps, and that in the rotating cells the single response takes place just when the shading of the flagellar base by the non-transparent cell wall begins. Halldal (1958) concluded that any part of the cell can function as a screen, provided that it reduces the intensity in the active spectral range by at least 10%, and that the stigma may have only an auxiliary function. However, not even intermittent shading by the cell can be regarded as a general mechanism of phototactic orientation, since flagellated forms exist which react photo-topotactically although they do not rotate while they swim (Stahl, 1880; Ringo, 1967). Thus *Euglena* has to be regarded as an exception rather than as a general model of phototactic orientation in flagellates. According to Halldal, the direction of movement under unilateral illumination is adjusted by an asymmetrical beating of the flagella when the photoreceptor is shaded by any part of the cell. This takes place until the flagellated end of the cells points to the light source, when the flagellar base is not further shaded.

In the dinoflagellate *Gyrodinium dorsum* the phototactic response consists of an initial stop response, followed by swimming in the direction of the stimulus beam (Hand *et al.*, 1967). Initial stop responses were also observed by Feinleib and Curry (1971b) in *Chlamydomonas* and Huth (1970a) in *Volvox*. Although these stop responses might be considered as photophobic responses (see page 62), they are an integral part of phototactic orientation. It seems that during the stop the organism determines the light direction by comparing the quantum fluxes in different regions of the cell, and then realigns itself so that its anterior part faces the light source. Obviously in this position the quantum fluxes at all measuring points are equal. So it seems that both the basic mechanisms for detecting light direction (see page 31) are realized in phototaxis, one in *Euglena* and the other in *Gyrodinium*.

(c) *Volvox*. In the colonies of *Volvox aureus* phototactic orientation results from unequal driving forces at opposite sides of the colony due to a stopping of flagellar beating of the cells on the irradiated flank (Gerisch, 1959; Huth, 1970a, b; Hand and Haupt, 1971).

(d) *Micrasterias*. The phototactic reactions of desmids, such as *Closterium* and *Micrasterias*, were studied by Stahl (1880), Klebs (1885) and Aderhold (1888). The phototactic behaviour of *Micrasterias* was reinvestigated by Bendix (1960) and Neuscheler (1967a, b). As shown by the latter, the *Micrasterias* cells display an oscillating movement, resulting in meander traces of mucilage (Fig. 9). Under unilateral illumination the periods and amplitudes of the oscillations decrease and the traces become straighter. Since the cells orientate their longitudinal axes parallel to the light beam and produce mucilage only at their rear pole, the traces become directed towards the light source. If the light direction is changed, the cells change their position until they face the light again. It is not yet clear where the photoreceptor is located and how the light direction is perceived.

(e) *Anabaena*. The phototactic orientation of some blue-green algae of the family Nostocaceae, such as *Anabaena* and *Cylindrospermum*, is also the result of an active steering mechanism (Drews, 1959). Under unilateral illumination most of the filaments form a U with its axis of symmetry oriented parallel to the light direction. When the light direction is changed, the bent "tip" begins to turn until the U is again directed to the light source. However, even approximately straight filaments display phototactic orientation. Since they do not rotate during movement, the bending is probably the result of a different behaviour of the illuminated and the shaded sides of the filament, perhaps comparable to phototropic responses.

2. *Change of autonomous rhythm of reversal*

Under diffuse light, blue-green algae of the family Oscillatoriaceae (Drews, 1957, 1959; Nultsch, 1961) and diatoms (Nultsch, 1956) display an alternating backward and forward movement without preferring any direction. On the onset of unilateral illumination the organisms do not change their movement direction actively. However, in individuals which are in a more or less parallel position to the light beam the movement toward the light source is prolonged while movement away is shortened resulting in positive photo-topotaxis. With negative reactions the situation is the opposite. In organisms which are oriented perpendicularly to the light beam no phototactic effect can be observed. However, since the movement of the organisms is never completely straight, all individuals of a population come, sooner or later, for a long or short time into a position more or less "parallel" to the direction of light.

The mechanism of the change of this autonomous rhythm under unilateral illumination is unknown. Some new findings, arising from partial illumination experiments in diatoms, will be discussed later (see page 64f).

E. Conclusions

Although many attempts have been made to elucidate the mechanism of photo-topotaxis neither the primary processes of light perception nor the secondary chain of stimulus transduction to the effector is fully understood for any organism. Some new ideas in this field are reported below (see page 67f). Because of the diversity of photo-topotactic reactions one might expect that more than one mechanism would exist. Not even a common photo-receptor pigment can be assumed, since it is improbable that flavins or flavoproteins can mediate phototactic responses in red light up to 730 nm, although Edmondson and Tollin (1971) have shown that the absorption of some flavoprotein complexes can extend to about 600 nm. From most action spectra we can conclude that no causal relationship exists between photo-topotaxis and photosynthesis because of the phototactic ineffectiveness of red light. This does not necessarily exclude indirect relations between these processes, e.g. a light triggered release and utilization of energy produced in photophosphorylation, as suggested by Diehn and Tollin (1967) and Tollin (1969) on the basis of inhibitor experiments. Surprisingly, in some *Phormidium* species light absorbed by biliproteins is active, but red light absorbed by chlorophyll *a* is ineffective so that not even in these organisms photosynthesis seems to be involved. The only exception seems to be *Anabaena variabilis*. In its action spectrum the participation of the photosynthetic apparatus in mediating photo-topotactic stimuli is indicated by maxima around 430 and 670 nm. That a directional cue and not an intensity difference is perceived, was shown by Buder (1917) using a convergent light beam. Under these experimental conditions an organism which reacts positively in a parallel or divergent light beam also displays positive reactions although the illuminance decreases on its way towards the light source. This was confirmed by Halldal (1958) in *Platymonas*.

III. Photo-phobotaxis

As mentioned above, photo-phobotactic reactions are caused by temporal changes of irradiance, dI/dt. The stimulus must exceed a distinct threshold value (i.e. a minimum change in irradiance ΔI from I_0 to $I_0 \pm \Delta I$). The ratio $\Delta I/I_0$ is called the discrimination threshold (Clayton, 1959). If the decrease—ΔI equals I_0, in other words, if the organisms are completely darkened, the least I_0 perceived by the cells is called the absolute or zero threshold. Some organisms react positively in dim light and negatively in bright light. If so, the intensity ranges of positive and negative reactions are separated by a smaller or broader intensity range, in which the cells do not respond at all. This is called,

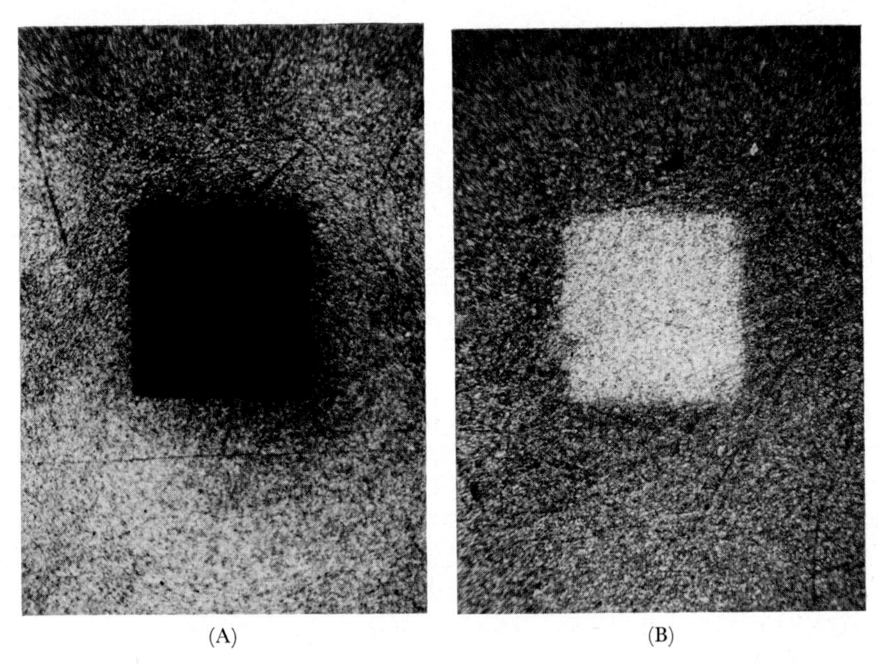

(A) (B)

FIG. 13. Positive (A) and negative (B) photo-phobotaxis demonstrated with the light trap method. Photographs of the light fields a ter removing the slits (courtesy of (Dr. D. P. Haeder.

as in photo-topotaxis, the indifference zone or inversion intensity (see page 42). However, there are also organisms in which only positive reactions have been observed. To be perceived, the "new" irradiance must be present for a minimum time, called the presentation time, while the period of time between the change of the irradiance and the motor response is called reaction time. If a second change of intensity follows the first one immediately, the cell does not respond at all (refractory period). Threshold intensities as well as presentation and reaction times depend on internal and external factors. As shown in the introduction, a stimulus (i.e. a change of intensity in time) is also perceived by an organism when it quickly traverses the border between two fields of different light intensities. Therefore, positively reacting organisms accumulate in Engelmann's "light trap", i.e. a light field projected onto a preparation of organisms kept in the dark or illuminated with lower intensities. The organisms are prevented from leaving it but not from entering it. Negatively reacting organisms can leave the light field but are not able to enter it, so that the field becomes empty (Fig. 13).

A. *Methods of study*

As with photo-topotaxis, population methods and individual cell methods can be used.

1. *Population methods*

All population methods for measuring photophobotactic reactions use the principle of Engelmann's light trap. The methods differ as to

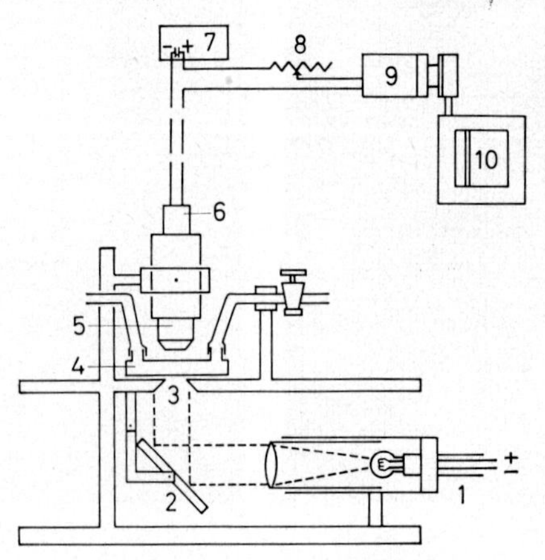

Fig. 14. System for measuring photo-phobotactic reactions of *Rhodospirillum rubrum*. 1. Tungsten lamp, 2. mirror, 3. slit, 4. flow-through cuvette, 5. microscope objective, 6. photoresistor, 7. anode battery, 8. variable resistance, 9. galvanometer, 10. recorder (modified from Throm, 1968).

whether a dark or illuminated background is used and whether the density of accumulation is evaluated visually or photometrically, during the experiment, or at the end of it.

(a) *Single beam systems.* In these systems a light field is projected onto a preparation of organisms in the dark. Since with slowly moving organisms, such as blue-green algae and diatoms, the accumulations are stable for a couple of minutes, their optical density can be measured at the end of the experiment (Nultsch, 1962b, 1971). With quickly moving organisms, however, such as purple bacteria, the optical density must be measured during the experiment. In this case, the actinic light serves

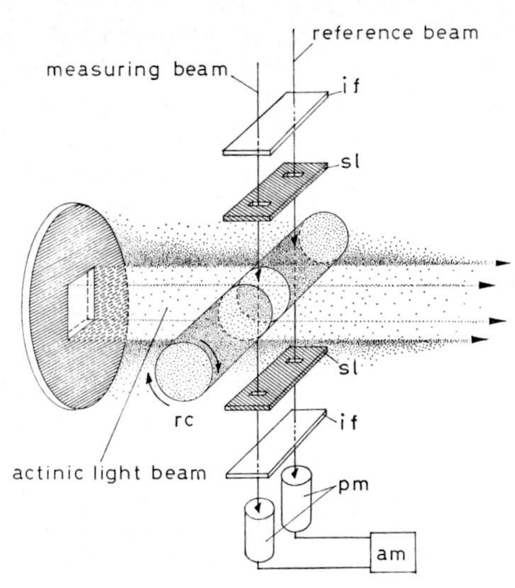

FIG. 15. Optical configuration of the phototaxigraph, am = operational amplifier, if = interference filter, pm = photomultipliers, rc = rotating cuvette, sl = slit.

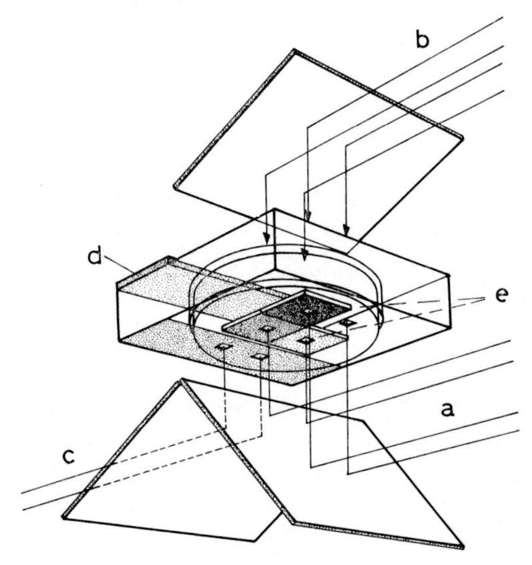

FIG. 16. The improved double irradiation technique used by Nultsch and Haeder (1974). a = actinic trap light, b = background light, c = white light control, d = dark background, e = neutral density filters.

simultaneously as the measuring light (Fig. 14), which is a considerable disadvantage. Single beam techniques were used also by Pohl (1948), Bruce and Pittendrigh (1956), Brinkmann (1966) and Bruce (1970), who investigated circadian rhythms in the photomotion of *Euglena* and *Chlamydomonas*.

(b) *Phototaxigraph.* In this system, devised by Lindes *et al.* (1965) and improved by Diehn (1969a), the actinic and monitoring light beams are separated. As shown in (Fig. 15), the measuring beam of 800 nm passes the actinic zone at right angles to the actinic beam, while the reference beam of the same wavelength passes the suspension outside of the actinic zone. Both are measured by photomultipliers coupled to an amplifier and recorder system. Background illumination can be used.

(c) *Double irradiation technique.* "Pure" photophobic responses can be measured in a system devised by Nultsch and Haeder (1970) for measuring the discrimination threshold of blue-green algae. Square light traps are projected from below onto a preparation of algae irradiated with a homogeneous background from above. Since the background illuminance is much stronger than the scattered light, no topotactic responses can occur. The optical density of the accumulations is measured at the end of the experiment. If different wavelengths are used as trap and background light, action spectra can be measured this way (Haeder and Nultsch, 1973; Haeder, 1974). An improved version of this technique was used (Fig. 16) to measure dose-response-curves at different trap and background wavelengths (Nultsch and Haeder, 1974).

(d) *Null method.* The intensities of two adjacent fields of different wavelengths are adjusted until they are equally attractive, enabling the relative effectiveness of the wavelengths to be determined. This method has also been used for white light experiments (Schrammeck, 1934; Manten, 1948; Schlegel, 1956).

Errors and misinterpretations are possible when light trap methods are used (Clayton, 1957, 1964). The following sources of error must be kept in mind:

(1) If organisms become motionless in the dark, they are able to leave a light trap, but cannot enter it if the surrounding field is dark. Negative photo-phobotaxis can thus be simulated by strong positive photokinesis. Conversely, in organisms which display negative photokinesis the movement can cease after entering a bright light field, simulating positive photo-phobotaxis.
(2) Since the photosynthetic activity of the organisms in a light field is higher than that of those outside, the oxygen concentration may increase and the CO_2 concentration decrease, producing chemical gradients in an enclosed system resulting in migrations into or out of the light field by positive or negative chemotaxis (see page 67).
(3) Accumulations in the light field can result from photo-topotactic reactions, since organisms may be attracted by light scattered laterally by organisms or other particles in the illuminated field (see page 36).

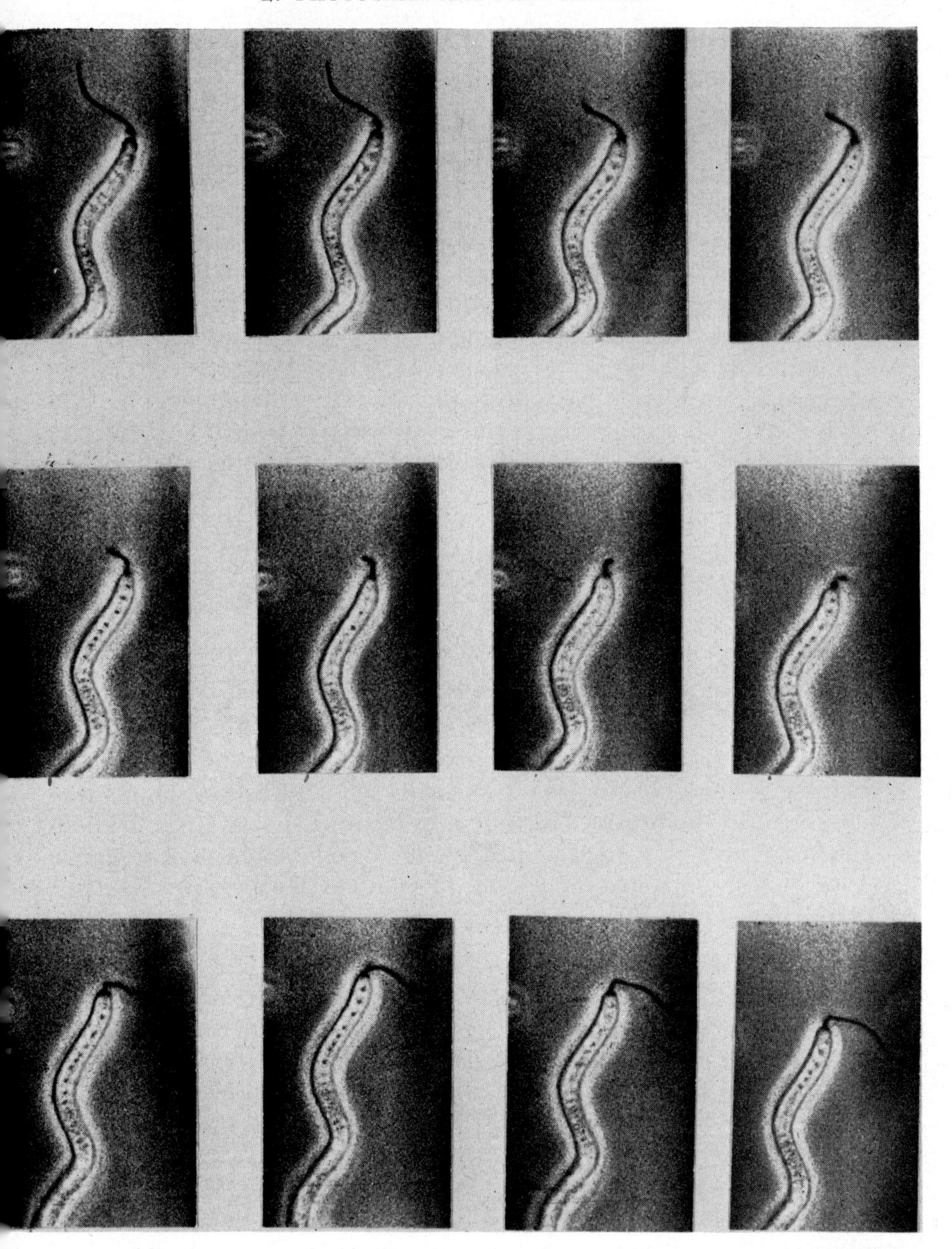

FIG. 17. "Snap over" of the flagella of *Thiospirillum jenense* during the photo-phobotactic response. The pictures are taken from the film "*Thiospirillum jenense,* Locomotion und phototaktisches Verhalten" with permission from N. Pfennig and the "Institut für den wissenschaftlichen Film", Göttingen.

2. *Individual cell methods*

Individual cell methods are indispensable for studying the behaviour of individuals during the reaction. However, they can also be used to estimate the reactions quantitatively. The most common method is the direct microscopic observation of single organisms. The behaviour of the cells can be studied, and also presentation and reaction times can be measured and threshold values can be evaluated by counting the numbers of responding cells. The method becomes more effective if combined with objective recording systems like those used for studying phototopotactic reactions, such as cinematography, flying spot scanning and computerized television (see section II). In order to localize light sensitive areas different parts of single cells can be illuminated by light fields of variable size with a special microscope in which the condenser is replaced by an achromatic objective. This partial illumination technique, originally developed by Buder (1915), has been improved by Nultsch and Wenderoth (1973).

B. Modes of response

The behaviour of organisms during photophobic responses varies from genus to genus and depends mainly on the morphology of the cell and on the mechanism of movement. The following types can be distinguished.

(a) *Gliding organisms.* In blue-green algae and diatoms the photophobic response is simply a stop which is usually followed by a resumption of movement in the opposite direction. Sometimes, however, individuals stop but resume their movement after a few seconds in the same direction as before and sometimes there is only a transient slowing down (Harder, 1920; Drews, 1959; Nultsch, 1956).

(b) *Spirilla.* With spirilla the reaction is also a light induced reversal, but it looks more "phobic" because of the higher swimming speed and the very short reaction time. In the bipolarly flagellated *Rhodospirillum rubrum*, which does not show any permanent polarity in movement, the flagella at both ends of the cell simultaneously reverse their beating direction, so that the rotation of the cell and hence the direction of movement is reversed. The behaviour of the unipolarly flagellated *Thiospirillum jenense* often resembles that of *Rhodospirillum*, since the flagellated pole can be the front as well as the rear. The reversal of the movement direction is accompanied by a "snap over" of the flagella. This is, however, not a passive deformation due to a change in the movement direction, but an active response, as observed by Buder (1915) and recently by Pfennig (Fig. 17). The snap over is the cause and not the result of the directional change. Under some conditions, one pole persists in acting as the anterior pole. In these circumstances a single light stimulus induces only a transient reversal of flagellum beat. The cell then moves forward again, but in a different direction due to the disturbance in flagellum beat (Buder, 1915).

(c) *Chromatium*. The purple bacterium *Chromatium* displays a strong polarity in movement. The flagellated pole is normally the rear pole. Upon a light stimulus, the bacterium performs a shock reaction *par excellence*. As described by Engelmann (1882), it swims back with increased speed a distance ten-to-twenty fold its cell length, as if recoiling from an invisible wall. As shown in Fig. 18, the directional change is the

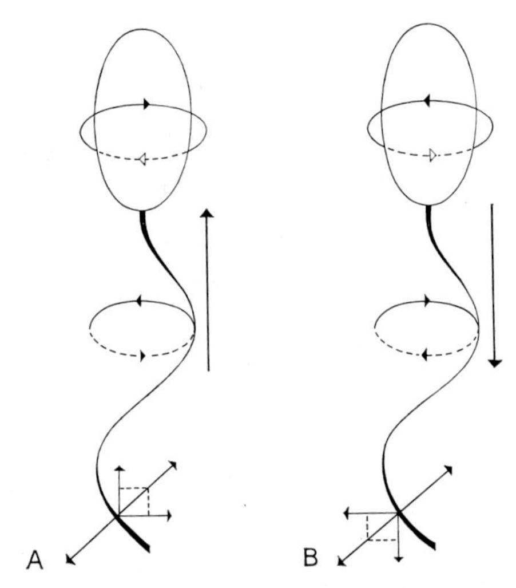

FIG. 18. Movement of *Chromatium*. A. Normal forward movement, B. backward movement during photo-phobotactic response (from Buder, 1915).

result of a change in the rotation direction of the flagellum, which acts as a propeller. After a "rest period" of a second or less, the bacterium resumes its normal forward movement. However, since during the rest period the orientation of its longitudinal axis has changed due to Brownian movement or to staggering after the reversal, the new direction diverges from the former one.

(d) *Flagellates*. In *Euglena*, in which the flagellum is inserted at the front, the photophobic response results in a sharp sideward beating of the flagellum to the ventral side, i.e. the side opposite to the stigma, so that the cell turns to the dorsal side. The turning impulse and, hence, the new movement direction depend quantitatively on the strength of the stimulus. Similar responses have been observed in other flagellates.

(e) *Volvox*. As mentioned above (p. 47) the topotactic orientation of *Volvox* colonies is the result of photophobic stop reactions of the cells on the irradiated flank. This stopping of flagellar beating also can be brought about by a sudden increase in light intensity (Huth, 1970a).

C. *Mechanism in prokaryotes*

Although there are differences in the methods of locomotion among photosynthetic prokaryotes showing photo-phobotaxis (i.e. purple bacteria and blue-green algae), the perception of the light stimulus seems to work through the same mechanism.

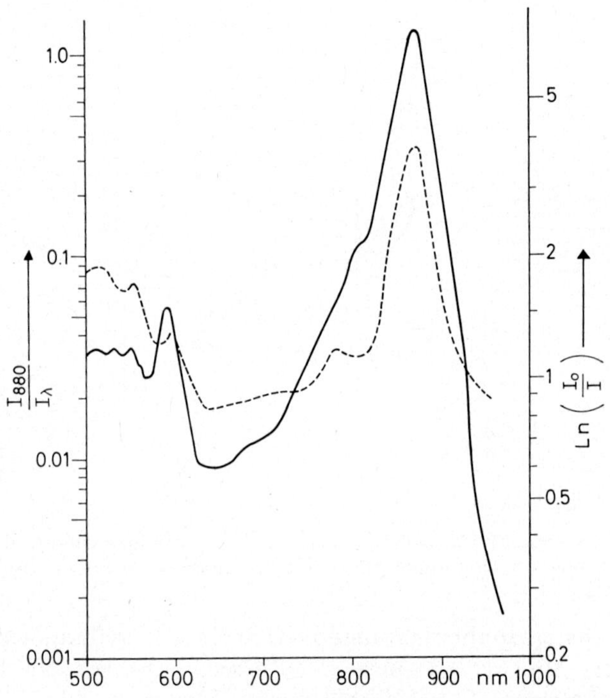

FIG. 19. Photo-phobotactic action spectrum (solid line) and *in vivo* absorption spectrum (dashed line) of *Rhodospirillum rubrum* (modified from Clayton, 1953a).

(a) *Action spectra.* Since the classical work of Engelmann (1883), Molisch (1907) and Buder (1919), we know that in purple bacteria the photo-phobotactically active light is absorbed by the photosynthetic pigments. This was confirmed later by Manten (1948), Thomas (1950), Duysens (1952) and Clayton (1953a), who measured action spectra of photo-phobotaxis (Fig. 19) and photosynthesis for *Chromatium* and *Rhodospirillum.* The spectra turned out to be essentially identical and coincided with the absorption spectra of the living cells. These findings led to the conclusion that in purple bacteria the photophobic responses are due to sudden changes in the rate of photosynthesis.

Action spectra studies in blue-green algae of the genus *Phormidium* gave similar results (Nultsch, 1962b, c). In general, the spectral ranges absorbed by the biliproteins C-phycocyanin and C-phycoerythrin are active. In addition, red light absorbed by chlorophyll *a* is active, too, while the effectiveness of blue light is far out of proportion to the absorption by the Soret band of chlorophyll *a* (Fig. 20). Since these action

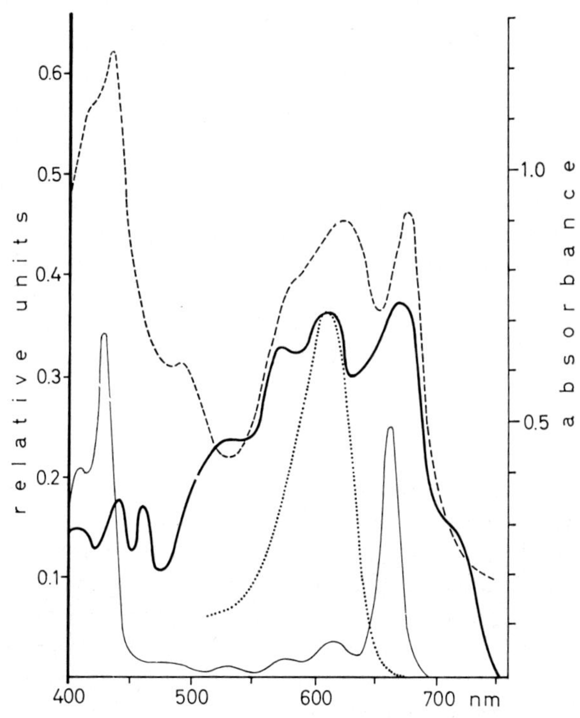

FIG. 20. Photo-phobotactic action spectrum (heavy solid line) and *in vivo* absorption spectrum (dashed line) of *Phormidium ambiguum*. Absorption spectra of chlorophyll *a* (fine solid line) and C-phycocyanin (dotted line). Abscissa: wavelength in nm; ordinate: photo-phobotactic reaction in relative units and absorbance (from Nultsch, 1962c).

spectra agree essentially with the photosynthetic ones measured by Duysens (1952), Haxo and Norris (1953) and Nultsch and Richter (1963) in some *Oscillatoria* and *Phormidium* species, one may conclude that in blue-green algae photo-phobotactic reactions are also mediated by the photosynthetic apparatus.

(b) *Inhibitor studies.* The coupling between photo-phobotaxis and photosynthesis was confirmed by Thomas and Nijenhuis (1950) and Clayton (1953a, b, c). The latter author concluded "that even when photosynthesis (as measured on steady state basis) is light saturated, a change in illumination could propel transient disturbances along the photosynthetic pathway and initiate a phobotactic response" (Clayton, 1964, p. 61). According to the hypothesis of Links (1955), these "disturbances" consist of changes of the ATP supply to the locomotor apparatus.

In order to decide these questions, the effect of photosynthesis inhibitors, uncouplers and redox-systems on the photo-phobotaxis of *Rhodospirillum rubrum* (Clayton, 1958; Throm, 1968) and *Phormidium uncinatum* (Nultsch, 1965, 1966, 1967, 1968, 1969, 1973a, 1974a; Nultsch and Jeeji-Bai, 1966) were investigated. In both organisms, inhibitors of photosynthetic electron transport decrease the photo-phobotactic reactivity, with the exception that DCMU does not inhibit photo-phobotaxis in *Rhodospirillum.* This is understandable, since purple bacteria lack photosystem II. The uncoupler experiments with blue-green algae gave equivocal results, since most of the uncouplers inhibit oxidative phosphorylation and, hence, movement in the dark. Consequently, in light trap experiments the organisms are prevented from entering the trap as a result of immobility, and no photo-phobotactic accumulations can occur. In *Rhodospirillum*, however, the photophobic response is not impaired by uncouplers, as has been shown by Throm (1968). In conclusion, the effect of photosynthesis inhibitors support the idea that photo-phobotactic reactions are coupled with photosynthetic electron transport, while the ineffectiveness of uncouplers is a convincing argument against the hypothesis of Links (1955).

In order to obtain more information about the coupling site between photo-phobotaxis and the electron transport chain, Nultsch (1968) has investigated the effect of artificial redox-systems, which, according to Witt *et al.* (1965, 1966), can function as electron acceptors for photosystems I and/or II. Redox-systems with midpoint potentials below $-0{\cdot}25$ V and between 0 and $+0{\cdot}1$ V inhibit photophobotaxis most strongly. These ranges of redox potentials correspond respectively with the first electron acceptors of photosystem I and II. However, since all redox-systems which penetrate the cytoplasmic membrane and which are able to trap electrons from the photosynthetic electron transport chain have some effect, the linkage cannot be restricted to these two sites of the electron transport chain.

(c) *Role of photosystem I and II.* Recently attempts were made to elucidate the role of photosystems I and II in photo-phobotaxis of *Phormidium uncinatum* by Haeder (1973, 1974), using the above mentioned double irradiation technique. He found that the phototactic reaction depends qualitatively and quantitatively on whether the background and trap wavelengths are absorbed by the same or by different photosystems. Nultsch and Haeder (1974) using a background wavelength of 563 nm absorbed mainly by C-phycoerythrin, and varying the trap wavelength, found a maximum of positive (exit) reaction (see below) around 700 nm, but negative (entrance) reactions below 440 and above 700 nm (Fig. 21), i.e. at wavelengths absorbed by photosystem I. On the other hand, if a background wavelength of 723 nm absorbed by photosystem I is used, maximum activity is observed in the range between 500 and 600 nm (Fig. 21), where light is mainly absorbed by C-phycoerythrin, i.e. the main accessory pigment of photosystem II (Amesz, 1973). With some

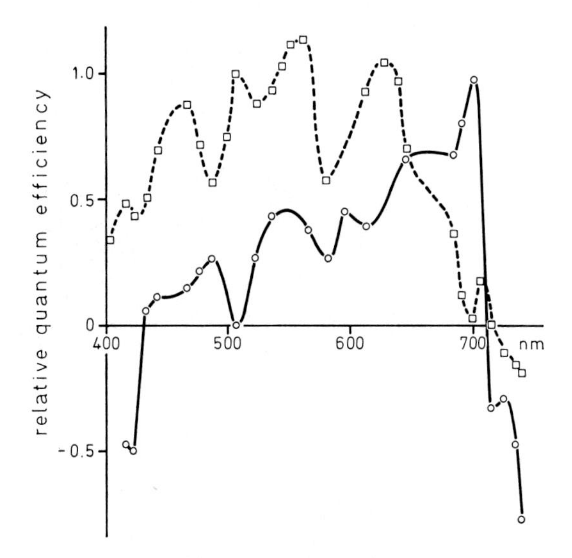

FIG. 21. Action spectra of phobic responses of *Phormidium uncinatum* with constant background wavelengths of 563 nm (circles and solid line) and of 723 nm (squares and dotted line). Abscissa: trap wavelength in nm; ordinate: relative quantum efficiency (after Nultsch and Haeder, 1974).

reservations the spectrum in a constant background of 723 nm can be regarded as an action spectrum of photosystem II, whereas a constant background wavelength of 563 nm roughly yields an action spectrum of photosystem I, in which even the emptying of the "trap" must be regarded as indicating strong phobotactic activity.

The emptying of the light trap resembles the pattern of negative photo-phobotaxis, previously unknown in this species. Therefore its wavelength dependence was investigated by Haeder and Nultsch (1973). The results are shown in Fig. 22, in which the wavelengths of trap and background light are plotted horizontally and the photo-phobotactic reaction values measured vertically. If the "inversion points" at which no reaction can be observed are connected with one another, a line is obtained which separates "positive" and "negative" reaction patterns. As can be seen, emptying of the light field occurs at all background wavelengths if far red or infrared is used as "trap" light. To explain these results, Haeder (1974) hypothesized that photo-phobotactic responses in *Phormidium uncinatum* are caused by a sudden decrease of an electron pool located in the electron transport chain between photosystem II and I

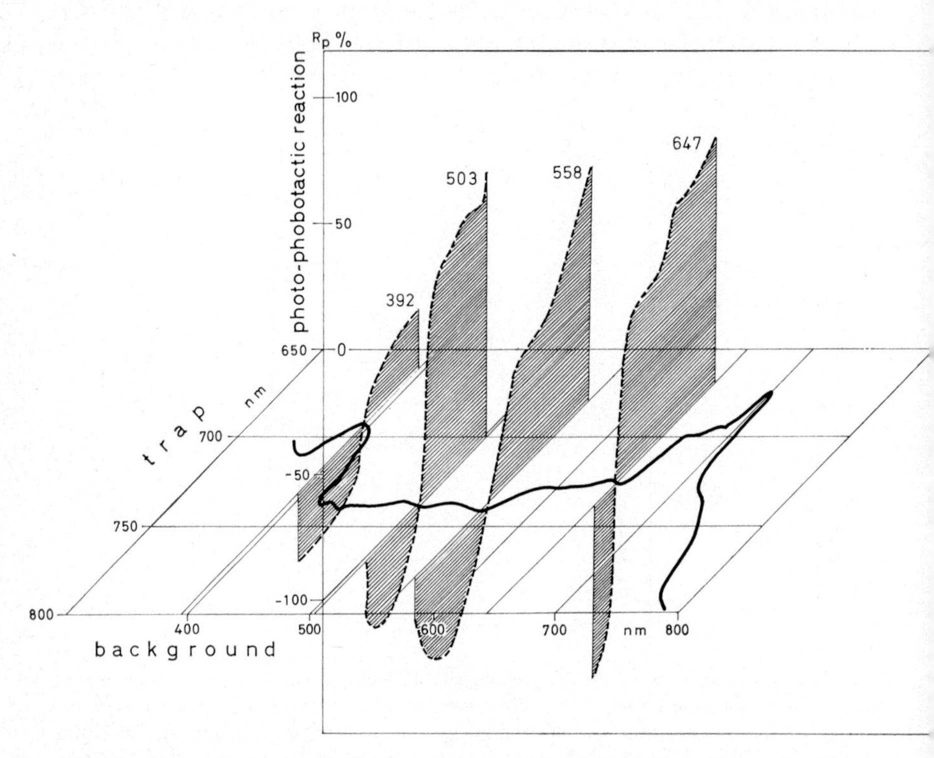

Fig. 22. Three dimensional diagram demonstrating the relation of "positive" (above the horizontal plane) and "negative" (below the horizontal plane) photo-phobotactic reaction values R_p in per cent (vertical axis) to trap and background wavelengths (horizontal axes). The irradiance of trap and background was constant 1 W m^{-2} each. For the sake of clarity, only the curves for the background wavelengths 392, 503, 558 and 647 nm are inserted. The solid line connects all "inversion points" between 342 and 781 nm (from Haeder and Nultsch, 1973).

(probably plastoquinone), which can be brought about either by interrupting the electron flow into the pool via photosystem II or by increasing the draining of electrons out of the pool via photosystem I. Consequently, whether the organisms are prevented from leaving the light field or from entering it depends on whether the absorption of quanta by photosystem II or I is greatest with the trap or the background wavelengths.

Thus, the emptying of the light field observed under these rather artificial experimental light conditions cannot be regarded as a result of true negative photo-phobotactic responses, which occur in other organisms at very high light intensities. Therefore, Nultsch and Haeder (1974) suggested the terms "exit reaction" and "entrance reaction" instead of "positive" and "negative" reactions, to avoid any misunderstanding.

(d) *Halobacterium halobium.* Recently, Hildebrand and Dencher (1974) found photophobic responses to occur in *Halobacterium halobium.* This interesting organism lacks a photosynthetic apparatus, but possesses the so-called purple membrane which consists of bacteriorhodopsin, a retinal-protein complex. This membrane functions as a light driven proton pump building up an electrochemical gradient across the cell membrane which apparently can be used by the cell for ATP synthesis (Oesterhelt and Stockenius, 1973; Hess and Oesterhelt, 1974).

Hildebrand and Dencher (1974) observed positive as well as negative photophobic responses. The action spectrum of positive phobotaxis shows a maximum around 565 nm which coincides with the absorption maximum of the isolated purple membrane. Although radiation of wavelengths below 460 nm which is also absorbed by the purple membrane (second maximum at 280 nm) was ineffective, bacteriorhodopsin seems to be the photoreceptor of this reaction. This is supported by the observation that cells of young cultures which lack a purple membrane do not show positive photophobic responses.

The action spectrum of the negative photophobic response is quite different. It displays maxima at 280 and 370 nm, a minimum at 310 nm and some smaller peaks between 400 and 500 nm. Radiation above 530 nm is ineffective. At shorter wavelengths the action spectrum resembles the absorption spectrum of an isolated retinylidene protein. As this pigment does not show absorption maxima between 400 and 500 nm, the authors consider that other chromoproteins must also act as photoreceptors. This second pigment system responsible for the negative photophobic responses is 15 times as sensitive as that for positive responses.

D. *Mechanisms in eukaryotic micro-organisms*

In the past it seemed that the mechanism of photo-phobotaxis in eukaryotic organisms differs principally from the one in prokaryotes in that perception is not mediated by the photosynthetic pigments (see Nultsch, 1973a, b, 1974a, b). Very recent investigations of Wenderoth (1975) and Nultsch and Schuchart (unpublished), however, gave evidence that two different photo-phobotactic mechanisms exist in eukaryotic organisms: the short wavelength type and the photosynthetic type, the latter being perhaps similar to that of prokaryotes. Most representative of the short wavelength type is *Euglena.*

(a) *Euglena.* The spectral sensitivity of negative photo-phobotaxis was investigated by Bünning and Schneiderhöhn (1956) and Gössel (1957). Using colourless strains with and without a stigma as well as green strains of *Euglena gracilis*, they found maximum activity between 400 and 420 nm. The strains differed in their sensitivity to light between 420 and 550 nm. Diehn (1969a, b) measured action spectra of positive and negative photo-phobotaxis of another *Euglena* strain using the phototaxigraph. The action spectrum of positive photo-phobotaxis resembles that of positive photo-topotaxis (Fig. 11) with peaks at 375 and 480 nm. The action spectrum of negative photo-phobotaxis, which is composed of two spectra obtained with polarized light the plane of which was parallel to the long axis of the cells in the one case and perpendicular to it in the other, shows two sharp peaks at 450 and 480 nm, a broad peak at 365 nm and a minor peak at 412 nm. Diehn suggested that the spectrum of negative photo-phobotaxis which can be considered as the absorption spectrum of the photoreceptor indicates a flavin as photoreceptor. Froehlich and Diehn (1974) have shown that the negative photophobic response of *Euglena* and a photoelectric effect induced in a lipid bilayer membrane containing a flavin display the same light intensity dependence.

(b) *Other flagellates.* In other flagellates photophobic responses have been little studied (Haupt, 1959). Stavis and Hirschberg (1973) investigated phototaxis of *Chlamydomonas reinhardtii*. However, they used a phototaxigraph-like system and made no attempt to separate topotaxis and phobotaxis. The initial stop response of the dinoflagellate *Gyrodinium dorsum* which might also be considered as a photophobic response (see page 46) is sensitive to blue light, but the maxima of the action spectrum which occur at 470 and 280 nm after red light (620 nm) irradiation, shift to 490 and 300–310 nm after irradiation with far red (Forward, 1970, 1973; Forward and Davenport, 1968). These findings are interpreted as indicating the existence of a two pigment system, in which a carotenoprotein functions as the photoreceptor pigment of the stop response, while phytochrome regulates its sensitivity. The shift in the action spectrum is explained as a *cis* to *trans* change of the carotenoid following phytochrome conversion.

(c) *Diatoms.* An action spectrum of photo-phobotaxis of the diatom *Nitzschia communis* was measured by Nultsch (1971). It resembles roughly that of positive photo-topotaxis. Differences were mainly found in the u.v. which is not as active in phobotaxis as in topotaxis. Neither the shape of the action spectrum nor the position of the peaks is sufficient to determine whether the photoreceptor is a carotenoid, flavin or other yellow pigment.

It must be mentioned, however, that Heidingsfeld (1943) with *Navicula radiosa* and Nultsch (1956) with some other species have observed accumulations in and migrations out of red light fields. Based on his experiments with *Nitzschia communis* Nultsch (1971) suggested that they are the result of aerotactic and photokinetic effects, respectively. Most recently, Wenderoth (1975) observed accumulations with sharp edges in red light fields with *Navicula peregrina* and measured the action spectrum (Fig. 23) using the single beam population method (see p. 50).

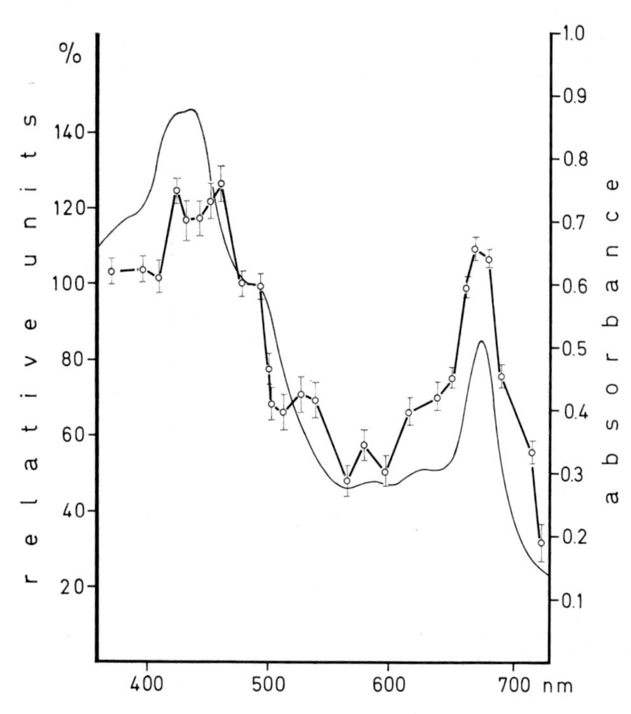

FIG. 23. Photo-phobotactic action spectrum (heavy line) and absorption spectrum (fine line) of the diatom *Navicula peregrina*. Abscissa: wavelength in nm; ordinate: photophobic response in relative units and absorbance (modified from Wenderoth, 1975).

The shape of the spectrum between 370 and 560 nm resembles that of *Nitzschia communis*. The red maximum at 669 nm coincides with the *in vivo* absorption maximum of chlorophyll *a*. Since Wenderoth was able to induce photophobic responses also in single cells by partial irradiation (see p. 54) with red light, there is no doubt that the accumulations of *Navicula peregrina* in red light fields are the result of true photophobotactic responses. Although *Navicula peregrina* is a eukaryote, correlations between photo-phobotaxis and photosynthesis seem to exist also in this organism, and it is probable that the red light responses of the other diatoms observed by Heidingsfeld (1943) and Nultsch (1956, 1971) are at least in part due to true photophobic responses, but are superimposed by aerotaxis and/or photokinesis. This will be clarified in further investigations.

Because of the similarity of the action spectra of topotaxis and phobotaxis in *Nitzschia communis*, the question has been raised whether both reactions in the short wavelength range are governed by the same photoreceptor. This idea is supported by the results of partial illumination studies in *Navicula peregrina* (Nultsch and Wenderoth, 1973). In these experiments, light fields of a constant size, the diameter being 25% of the cell length, were projected onto different parts of a *Navicula* cell moving in the dark (Fig. 24). If the middle of the cell is irradiated (C),

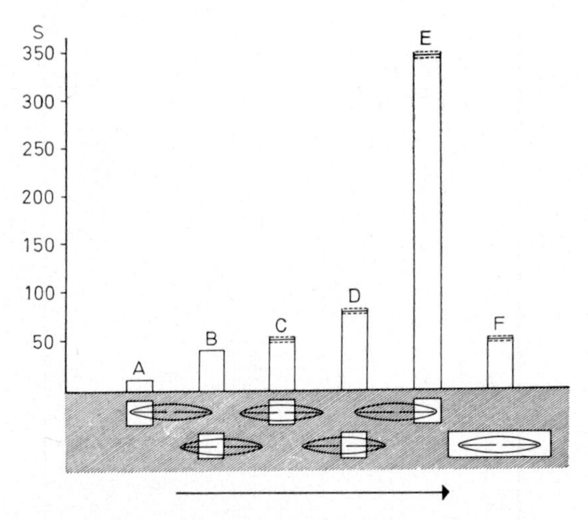

FIG. 24. Partial illumination experiments in *Navicula peregrina*. Arrow: movement direction. Ordinate: reaction time to reversal of movement in seconds. The diagram at the base of the figure indicates the part of the cell that is illuminated. In F the whole cell is illuminated for measuring the period lengths of the autonomous rhythm. For further explanation see text.

the reaction time equals the period of the autonomous rhythm. The more the light field is shifted to the rear pole (B, A), the shorter is the reaction time, reaching 10 seconds when the rear pole itself is irradiated (A). This equals the reaction time measured in light trap experiments, i.e. the time between the beginning of shading of the front pole of cells leaving the light trap, and the induced stop. On the other hand, the reaction time exceeds the period length of the autonomous rhythm as the light field comes closer to the front pole (D, E). This way it is possible to "draw" the cells, which follow the light spot over a long distance, and to change the autonomous rhythm of forward and backward movement.

In photo-topotaxis a delay of the reversal resulting in a prolonged movement toward the light source is caused by the signal: "front half receives more quanta per time unit than the rear half". *Vice versa*, a premature reversal resulting in a shortened movement away from the light source is induced by the signal "front pole receives less quanta per time unit than the rear half". Although positive photo-phobotactic responses at a light trap border can be also induced by the latter signal, they are normally caused by the signal "less quanta per time unit than before" measured at the front pole. This means that in photo-topotactic

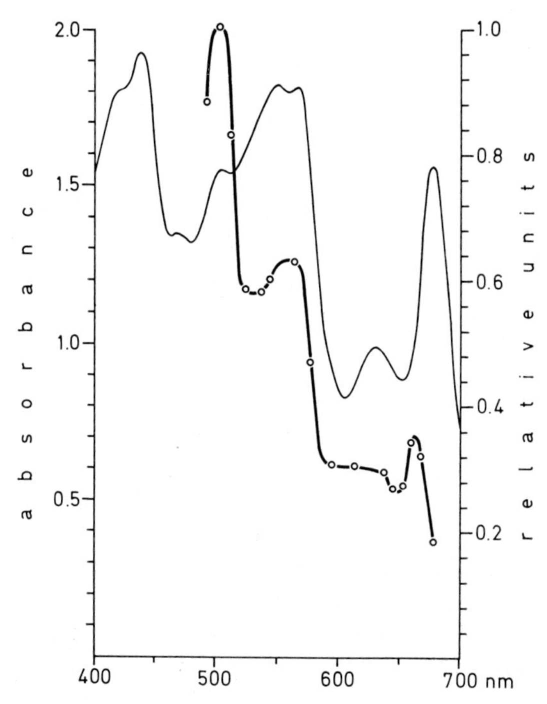

FIG. 25. Photo-phobotactic action spectrum between 493 and 692 nm (heavy line) and absorption spectrum (fine line) of the red alga *Porphyridium cruentum*. Abscissa: wavelength in nm; ordinate: photo-phobic response in relative units and absorbance (Nultsch and Schuchart, unpublished).

reactions the light absorbed in the front and rear of the cell at the same time is compared, i.e. a spatial comparison, whereas in photo-phobotactic reactions the light absorbed in a single region of the cell, the front pole, at different times is compared, i.e. a temporal comparison. In both cases the same photoreceptor could be used. The ability of the cells to compare spatial as well as temporal differences in irradiance has been confirmed most recently by short time partial illumination experiments (Wenderoth, 1975).

(d) *Porphyridium cruentum.* Since *Navicula peregrina* is the first eukaryotic organism in which true positive photo-phobic responses in red light could clearly be demonstrated, indicating correlations between photo-phobotaxis and photosynthesis, it was of great interest to investigate other eukaryotic forms. As flagellates for the above mentioned reasons seem not to be promising objects, the red alga *Porphyridium cruentum* which displays gliding movement was used. In preliminary experiments (Nultsch and Schuchart, unpublished) accumulations in red fields have been observed. The spectrum so far measured (Fig. 25) shows a main maximum at about 500 nm and two smaller ones around 560 and 665 nm. The first one seems to be mainly the result of the photo-topotactic action of laterally scattered light (its wavelength coincides roughly with the maximum of the photo-topotactic action spectrum). The maxima around 560 and 670 nm, however, coinciding with the main absorption band of B-phycoerythrin and the red maximum of chlorophyll *a*, respectively, are the result of true photo-phobic responses because of the complete photo-topotactic ineffectiveness of radiation above 550 nm. This points to correlations between photo-phobotaxis and photosynthesis also in this alga.

E. Conclusions

From action spectra it can be concluded that two main types of photo-phobotactic mechanisms exist: 1. the photosynthetic type, in which the phobic reactions are governed by the photosynthetic apparatus, and 2. the short wavelength type, in which yellow pigments function as photoreceptors. A third type, the *Halobacterium* type, differs from the photosynthetic one in lacking a photosynthetic apparatus but resembles it in so much as the light energy conversion apparatus is also used for triggering the positive photophobic responses. This enables individuals of both types to stay in light conditions which are favourable for energy conversion.

Since in earlier investigations the photosynthetic type was found only in purple bacteria and blue-green algae, Nultsch (1973a, b; 1974a, b) suggested it was restricted to prokaryotes. However, from the action spectra of *Navicula peregrina* and *Porphyridium cruentum* it must be concluded that correlations between photo-phobotaxis and photosynthesis also exist in eukaryotic organisms.

As mentioned above, photo-phobotactic responses are, at least in purple bacteria and blue-green algae, triggered by the photosynthetic electron transport. This may be valid for all organisms of photosynthetic type. The mechanism of stimulus transduction to the locomotor apparatus, however, is an open question. Clayton (1959, p. 384) suggested that photophobic responses in purple bacteria are "mediated through the development of an excitatory state which is transmitted to the locomotor areas, causing a co-ordinated motor response". Nultsch

(1970, 1973a, b) has discussed a similar mechanism in the case of blue-green algae. Since correlations between photosynthesis and bioelectric potentials have been demonstrated in several algal species (Schilde, 1966; Throm, 1970, 1971), one could imagine that changes in the redox state of the above mentioned electron pool are transformed into membrane potential changes, which are conducted by the cytoplasmic membrane to the locomotor apparatus. This can easily be understood in prokaryotic cells in which the photosynthetic membranes originate from the cytoplasmic membrane and are directly connected with it. In eukaryotic cells the situation is more complicated because the thylakoids are encapsulated by the chloroplast envelope which prevents the immediate transmission of electric potential changes to the locomotor apparatus by the cytoplasmic membrane.

The short wavelength type is represented by the flagellates investigated so far. From the action spectra published we know that only u.v., violet, blue and blue-green light is active, although for the above mentioned methodological reasons we are not sure if in all studies "pure" photo-phobotactic reactions have been measured. Accumulations of *Euglena* cells in red light fields reported by Wolken and Shin (1958) and by Checcucci *et al.* (1974) may be the result of positive chemotaxis to oxygen. Even the effect of DCMU on the phototaxis of *Euglena* in white light, observed in the phototaxigraph by Diehn and Tollin (1967), could at least in part be explained this way. Checcucci *et al.* (1974), who worked with a modified version of this apparatus, were able to inhibit accumulations in red light by adding 10^{-7}–10^{-4} M CMU, which did not affect accumulation in blue-green light. Stavis and Hirschberg (1973), who used a similar apparatus but *Chlamydomonas* as test organism, did not find any effect of DCMU on phototaxis. It would hence be of interest to determine whether *Chlamydomonas* reacts aerotactically. Also experiments with other inhibitors yielded contradictory results in the two organisms. For example, azide inhibited motility and phototaxis in *Euglena* equally, but only phototaxis in *Chlamydomonas* (Stavis, 1974). Conversely, uncouplers of the CCCP type completely inhibited motility and phototaxis in *Chlamydomonas*, but only phototaxis in *Euglena*. Interpretation is rendered more difficult because topo- and phobotaxis are not clearly distinguished by the phototaxigraph technique. The observation that negative photophobic responses in *Euglena* in a reducing environment can be caused by red light (620–680 nm) of high intensity (Creutz and Diehn, 1972, see Diehn, 1973) should be noted.

Since most of the organisms use the same light absorbing system for photosynthesis and photo-phobotaxis, it is difficult to understand why only the flagellates seem to have developed a special photoreceptor for governing photo-phobotactic responses, e.g. the paraflagellar swelling in *Euglena*. If it is lost, no photophobic responses occur (Gössel, 1957). The photoreceptor, which is located near the flagellar base, delivers the stimulus to the flagellum, causing a phobic response. The mode of stimulus transduction is unknown but various models have been proposed. Diehn and Tollin (1967) and Diehn (1973) suggest that the "motor" (flagellum) is energized by either oxidative or non-cyclic photosynthetic phosphorylation *via* one or two effectors which receive the signal from the receptor. Tollin (1969) suggested that the flagellum would be provided with a pulse of ATP this way, resulting in a sharp arrhythmic contraction and causing a phototactic response. Bovee and Jahn (1972) assumed the photoreceptor to be piezoelectric, acting as a capacitor which discharges when the intensity of the impinging light is changed. The charge delivered, they suggest (p. 259), "augments the effect of ion movements along the flagellum, also augmenting the amplitude and force of the flagellar undulations and altering the position of the flagellum relative to the body and the direction of swimming". The "calcium current hypothesis" (Eckert, 1972) explains the reversal of ciliary beating in *Paramecium* by an increase in the intraciliary Ca^{2+} concentration due to an increase in Ca^{2+} conductance which is caused by a depolarization of the membrane. K^+ ions have an antagonistic effect. Since Halldal (1957) has found that the sense of the phototactic reactions is determined by the ratios of Mg^{2+}/Ca^{2+} and Ca^{2+}/K^+ ions, Eckert's hypothesis could be a promising model for the explanation of the mechanism of phototactic responses in flagellated cells. The above mechanisms can also explain the photo-topotactic reactions of flagellates if one assumes that they are brought about by a succession of photo-phobotactic responses.

IV. Photokinesis

The first detailed studies on photokinesis were carried out by Bolte (1920). She distinguished between positive and negative photokinesis and gave the following definitions: Negatively photokinetic organisms become motionless in the light, but continue to move in the dark until they either die or become motionless for other reasons. Positively photokinetic organisms are motile in the light, but become motionless in the dark after a time. Photokinetically indifferent organisms show no effect of light on motility.

Haupt (1959) has pointed out that the term photokinesis has been used by several authors in different senses. Contrary to Bolte's definition, most of them define positive photokinesis as an increase of the linear velocity by light and negative photokinesis as a light induced decrease of speed. Later, Nultsch (1962a) has shown that these definitions are equivocal in as much as the estimation of acceleration and deceleration depends on the reference point. In relation to the photokinetic optimum at which maximum speed is achieved, the effect of any lower (or higher) light intensity would be "negative". Nultsch therefore used the speed of movement in the dark as reference value. *Positive photokinesis* is then an *acceleration of linear velocity by light related to the velocity in the dark*, independently of whether or not the organisms continue to move in the dark. The photokinetic effect is *negative* if the *linear velocity in the light is lower than in the dark*. The initiation of movement by light in forms which become motionless in the dark must also be regarded as positive photokinesis, while negative photokinesis at high light intensities can result in immobility. Organisms which become motionless in the dark cannot show negative photokinesis, since the reference speed is zero.

A. *Methods of study*

Since photokinesis is an effect of light on the linear velocity of movement, it can be evaluated simply by speed measurements under defined light conditions. Either the distance traversed in a given time or the time needed to traverse a defined distance can be measured. This can be done directly or indirectly in either single cells or cell populations.

1. *Population methods*

The main advantage of population methods is that they give the average behaviour of a large number of cells.

(a) *Direct measurement of random motility.* An assay for measurement of random motility in bacteria was devised by Adler and Dahl (1967) and improved by Stavis and Hirschberg (1973). A capillary tube is filled with medium and then a small portion of a concentrated suspension of organisms is added at one end of the tube. The tube is held in a horizontal position under test conditions for some hours. At the end of the experiment the tube is broken into pieces and the number of organisms in each compartment evaluated either by plating the diluted suspension on agar and counting the colonies (Adler and Dahl, 1967) or by measuring the radioactivity if ^{14}C-labelled cells are used (Stavis and Hirschberg, 1973). The "motility coefficient" M, which has the dimension of square centimeters per hour, is then calculated using the following formula:

$$\ln c/c^0 = -1/2 \ln (\pi \mathrm{M}t) - x^2/(4\mathrm{M}t),$$

where c^0 is the initial number of organisms at the origin and c the number of organisms at a point x cm away from the origin after time t.

Nultsch (1962a, 1971) has described a similar method for measuring photokinesis in diatoms and blue-green algae. Agar plates are centrally inoculated with masses of algae which migrate over the surface of the plates during the following hours. After a given time, the diameter of the spreading area in the light (d_l) and in the dark (d_d) is measured. The photokinetic effect R_k is defined as the ratio d_l/d_d.

(b) *Doppler effect technique.* A new technique for measuring the linear and rotational velocity of *Euglena* was recently devised by Ascoli *et al.* (1971). The moving cells are aligned by means of an electromagnetic field at a radiofrequency of 10 MHz. A laser beam (6,327 Å) is scattered by the cells and received by a detector which lies in the plane of the laser. The shift of the frequencies of the scattered beams due to the Doppler effect serves as a measure for the linear velocity. For more details see Checcucci (1973).

(c) *Indirect measurement of relative speed differences.* If only relative speed differences and not the absolute linear velocity are of interest, the kinetics of phototactic reactions can be used to compare the speed of populations under different conditions. This can be done in both topotaxis and phobotaxis. In experiments with the purple bacterium *Rhodospirillum rubrum* Throm (1968) used differences in the slopes of photo-phobotactic recordings as a measure of the photokinetic effect.

2. *Individual cell methods*

The speed of single cells can be measured either directly or indirectly, i.e. by measuring their tracks. Moreover, the number of cells swimming through a known area in a given time can be counted. For cells motionless in the dark either the light intensity and the time of irradiation necessary for the initiation of movement or both can be measured.

(a) *Speed measurements with the microscope.* The speed of single organisms can be measured under defined light conditions, the time needed to traverse a given distance usually being measured. It is not possible to measure the speed of movement in darkness in this way, but approximate values can be obtained by measuring the speed at ineffective wavelengths or intensities. Since deviations due to individual variations are considerable, a large number of measurements must be averaged to obtain reproducible results. This is the main disadvantage of the method. Microscope measurements of linear velocity can also be done with video-recording systems (Ojakian and Katz, 1973; Schneider and Doetsch, 1974) or cinematography (Gray, 1955; Philipps, 1972). These methods are, however, time-consuming and expensive.

(b) *Measurements of tracks.* The linear velocity of individual cells can also be evaluated by measuring the length of tracks made in a given time. Feinleib and Curry (1967) devised a photomicrographic method for measuring the swimming rate of *Chlamydomonas*. They took photographs with a fixed exposure time and then measured the length of the tracks described by the swimming cells on the film (Fig. 26). One has to make sure that the cells have remained in focus during the exposure because the vertical component of swimming is a possible source of error. Similar photomicrographic techniques have been used by Gibbons and Gibbons (1972) and Kung (1971). Tracks recorded in the "flying spot scanning device" (see page 38) or in the video recording system described by Davenport *et al.* (1970) can also be measured. In the desmid *Micrasterias*

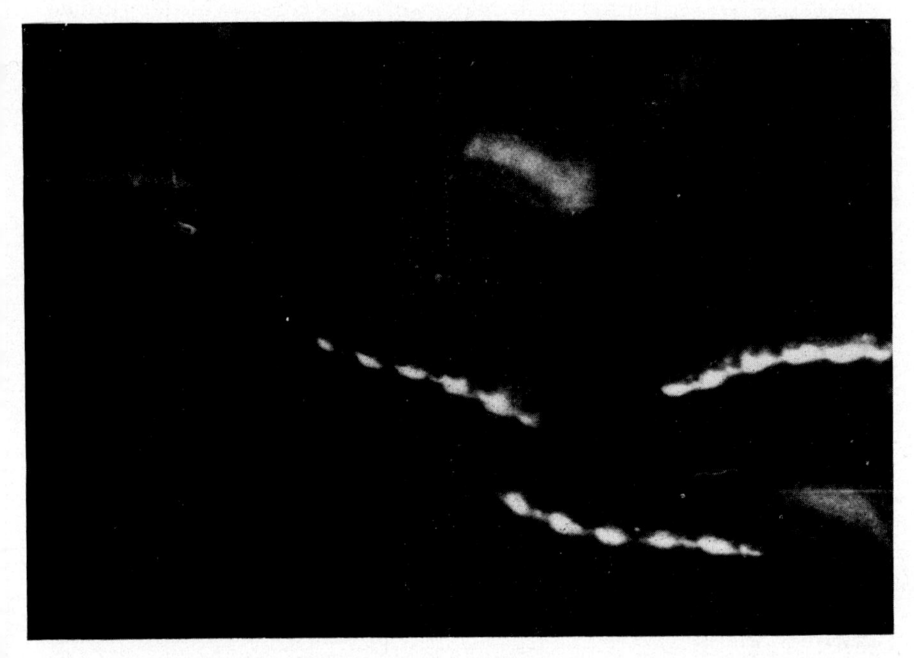

FIG. 26. One-fifth second dark field exposure showing swimming tracks of *Chlamydomonas* cells in dim red light plus a blue-green stimulus from the left side (courtesy of Dr. Mary Ella Feinleib).

denticulata, Neuscheler (1967b) evaluated the velocity of movement by measuring the length of the mucilage traces left by the creeping cells on the surface of the substrate in a given time (Fig. 9).

(c) *Cell counting methods.* The swimming speed of the organisms is not measured directly, but the number of cells swimming through a defined area in a given time is counted. Ojakian and Katz (1973) used a hemacytometer mounted on the stage of a microscope and counted the number of cells under phase-contrast or darkfield illumination. The swimming speed was calculated from the following formula:

$$V = \left(\frac{N}{t}\right) \cdot \frac{4}{nA}$$

where V is the average swimming speed, N is the number of cells passing through the plane area in the given time t, n is the cell density of the suspension, A is the area and 4 is a statistical factor. If the cells are confined to movement in a plane, π must be substituted for 4. With the technique of Rikmenspoel and van Herpen (1957) mentioned earlier (see page 40f.), either the number of cells passing the aperture in a given time can be counted or the linear velocity of single cells can be determined by measuring the duration of the flashes.

A method in which the photokinetic effect was quantified without speed measurements was used by Luntz (1931a, b, 1932). He evaluated the "Bewegungsschwelle" (motility threshold), the light intensity or light quantity ($I \times t$) which causes a resumption of movement in organisms which had become motionless in the dark.

B. *Intensity dependence*

In the investigations of Strasburger (1878), Engelmann (1882, 1883, 1888) and Bolte (1920), the effect of light on motility was investigated

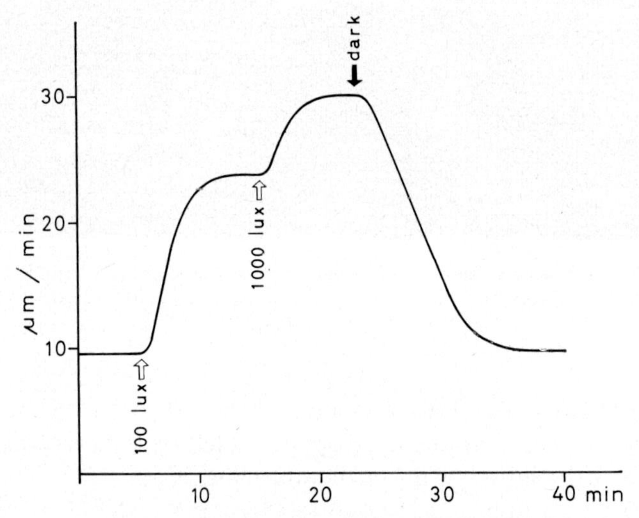

FIG. 27. Demonstration of the photokinetic effect by speed measurements in *Anabaena variabilis* (after Nultsch, 1974a).

only qualitatively. Luntz (1931a, b, 1932) evaluated the "motility threshold" in *Volvox*, *Eudorina* and *Chlamydomonas*. He found the photokinetic effect to be a function of light quantity ($I \times t$) or, under continuous illumination, of light intensity. Holmes (1903), Oltmanns (1917) and Mast (1926) found a relationship between linear velocity and light intensity in *Volvox*, but Mainx (1929) failed to demonstrate any effect of light on the speed of movement in this genus or in *Eudorina* and some other forms, nor were Haxo and Clendenning (1953) able to do so with *Ulva* gametes. Feinleib and Curry (1967) using their photomicrographic method mentioned above did not find a significant change of the swimming rate over a 10,000 fold range of light intensities in *Chlamydomonas*.

Wolken and Shin (1958) investigated the effect of polarized and non-polarized white light on the swimming rate of *Euglena gracilis*. At lower intensities they observed an acceleration of movement with

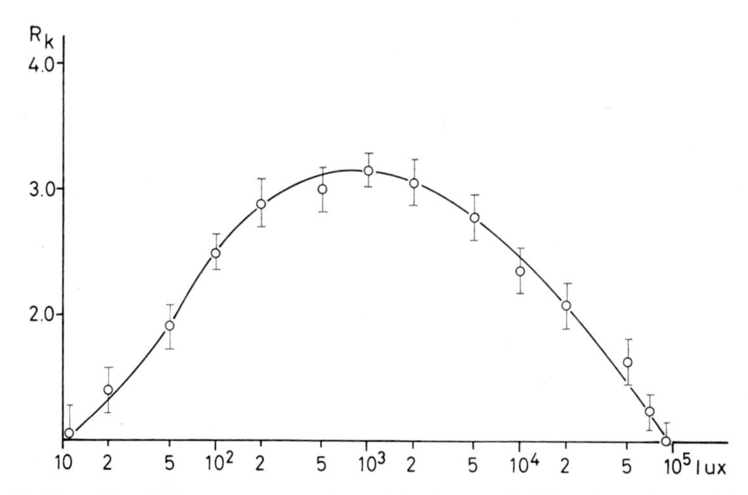

FIG. 28 Photokinetic dose-response-curve of *Anabaena variabilis* in white light. Abscissa: light-intensity in lux; ordinate: R_k = photokinetic effect in relative units (after Nultsch, 1974a).

increasing illuminance. The maximum speed of 0·16 mm/sec was reached at about 400 lux. This rather low saturation level may explain why photokinetic effects in *Euglena*, *Chlamydomonas* and other organisms could not be demonstrated by many authors (see Haupt, 1959).

Photokinesis in blue-green algae was investigated by Nultsch (1962a, c, 1974a). Fig. 27 shows results of microscope speed measurements in *Anabaena variabilis*. The average linear velocity in the dark, measured under ineffective blue light of wavelength <450 nm (see action spectrum, Fig. 31), was 9·5 μm/min. It was increased to 23·5 μm/min by white light of 100 lux, while at 1,000 lux the maximum speed of 30 μm/min was reached. Subsequent darkness decreased the speed to the original value. The dose-response-curves of photokinesis in the blue-green algae investigated are typical optimum curves (Fig. 28). In *Anabaena*, above 10 lux the linear velocity is a function of light intensity, up to photokinetic optimum at 1,000 lux. A further increase of illuminance reduces the photokinetic effect. At 9×10^4 lux speed equals the dark value, i.e. no photokinetic effect is detectable. Beyond this intensity, the velocity is lower than in the dark, as a result of negative photokinesis.

C. *Action spectra*

Action spectra have been measured in purple bacteria, blue-green algae, flagellates, diatoms and red algae. Most of them indicate correlations between photokinesis and photosynthesis.

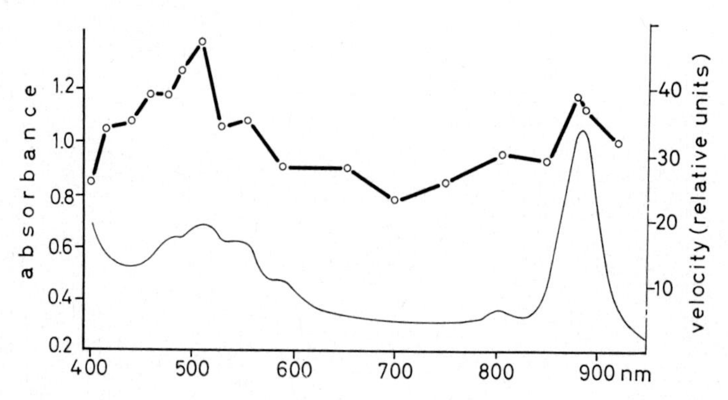

FIG. 29. Photokinetic action spectrum (open circles and heavy line) and *in vivo* absorption spectrum (fine line) of *Rhodospirillum rubrum* (after Throm, 1968).

(a) *Purple bacteria.* Engelmann (1883) investigated the photokinetic effect of coloured light in some purple bacteria. He found maximum activity in the infrared between 800 and 900 nm, minor but significant activity in the orange, while violet, blue and red had little or no effect.

An action spectrum of photokinesis in *Rhodospirillum rubrum* was measured by Throm (1968). Since the *Rhodospirilla*, unlike the bacteria investigated by Engelmann, were motile in the dark, Throm measured the light-induced increase of speed and not the effectiveness in causing motility. The differences in the effectiveness of the various spectral ranges are therefore not as striking as in Engelmann's experiments (Fig. 29). Two maxima of activity exist: one at about 890 nm due to bacteriochlorophyll *a* absorption and one near 510 nm, indicating the participation of spirilloxanthin in the absorption of the photokinetically active light. If we compare this spectrum with both the absorption spectrum of the cells and the action spectrum of photosynthesis (Clayton, 1953a; Duysens, 1952; Manten, 1948; Thomas, 1950), we find the relative height of the two peaks at shorter and longer wavelengths to be inverse, pointing to a stronger participation of carotenoids in photokinesis than in photosynthesis. Nevertheless, the action spectrum indicates that the photokinetically active light is absorbed by the photosynthetic pigments.

(b) *Blue-green algae.* Photokinetic action spectra for several blue-green algae were measured by Nultsch (1962a, c) and Nultsch and Hellmann (1972). In the three *Phormidium* species investigated light absorbed by chlorophyll *a* is highly active, while the ranges absorbed by carotenoids and biliproteins are not as active as the light absorption would suggest (Fig. 30). This finding indicates that photokinesis in the *Phormidium*

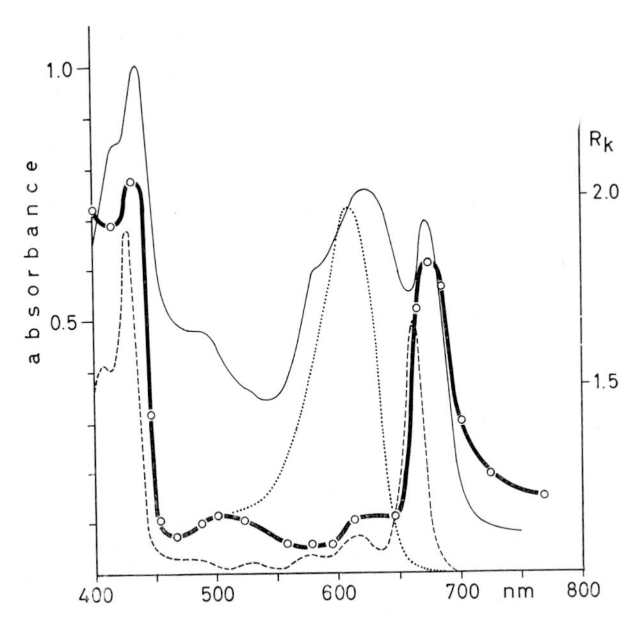

FIG. 30. Photokinetic action spectrum (circles and heavy solid line) and *in vivo* absorption spectrum (solid fine line) of *Phormidium ambiguum*. Absorption spectra of chlorophyll *a* (dashed line) and C-phycocyanin (dotted line). R_K = photokinetic effect in relative units (modified after Nultsch and Hellmann, 1972).

species is coupled with a process driven by photosystem I, probably cyclic photophosphorylation (Nultsch, 1970).

However, this is not general. Nultsch and Hellmann (1972) found that the photokinetic action spectrum of *Anabaena variabilis* is quite different (Fig. 31). Its maximum coincides with the absorption maximum of C-phycocyanin. Red light absorbed by chlorophyll *a* is also active, although no distinct peak or shoulder can be detected, while blue light absorbed by the Soret band is ineffective. Thus the photokinetic action spectrum of *Anabaena* resembles the photosynthetic action spectra measured in some blue-green algae (cf. Haxo, 1960). Consequently, the photokinetic effect in *Anabaena* must be ascribed to a process driven by both photosystems I and II, i.e. either by non-cyclic or by pseudocyclic phosphorylation (Nultsch and Hellmann, 1972).

Since the genus *Phormidium* belongs to the family Oscillatoriaceae and the genus *Anabaena* to the Nostocaceae, it was proposed that the different photokinesis types were characteristics of the families. Therefore, representatives of other genera of these families were studied by Nultsch

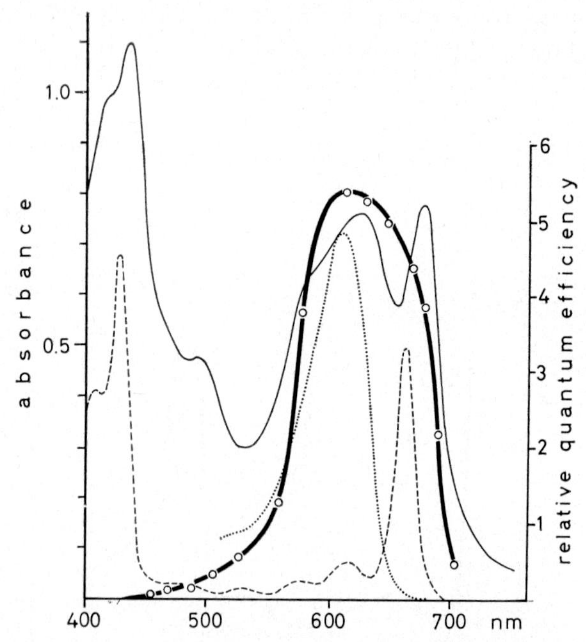

FIG. 31. Photokinetic action spectrum (circles and heavy solid line) and *in vivo* absorption spectrum (solid fine line) of *Anabaena variabilis*. Absorption spectra of chlorophyll *a* (dashed line) and C-phycocyanin (dotted line). (Modified after Nultsch and Hellmann, 1972).

(1973a, 1974a): the Oscillatoriaceae *Spirulina subsalsa*, *Oscillatoria tenuis* and *Pseudanabaena catenata* and the Nostocaceae *Nostoc. muscorum* and *Nostoc. calcicola*. These investigations confirmed the dominant role of photosystem I in photokinesis of Oscillatoriaceae, but revealed that biliproteins and carotenoids can also be active. On the other hand, the spectral sensitivity of both the *Nostoc.* species resembles that of *Anabaena*, except that the carotenoid range is more effective. From these results Nultsch concluded, that two main types of photokinesis exist: the *Phormidium* type, which is characterized by a strong photosystem I activity, and the *Anabaena* type, in which both photosystems are active. There are, however, "mixed" types. If we assume that in the "pure" types, cyclic photophosphorylation on the one hand and non-cyclic or pseudocyclic phosphorylation on the other serve as energy sources of photokinesis, we must conclude that in the mixed types both photophosphorylation processes supply energy to the locomotor apparatus.

(c) *Flagellates.* Luntz (1931a) presented rough photokinetic action spectra of flagellates from different taxonomic groups, such as *Eudorina*, *Pandorina* and *Chlamydomonas*, using the motility threshold (see page 72) as a criterion of the positive photokinetic effect. In all these species he found maximum activity around 492 nm, a result consistent with the phototaxis maximum of these organisms.

The action spectrum of positive photokinesis of *Euglena gracilis* was investigated by Wolken and Shin (1958), who measured the rate of

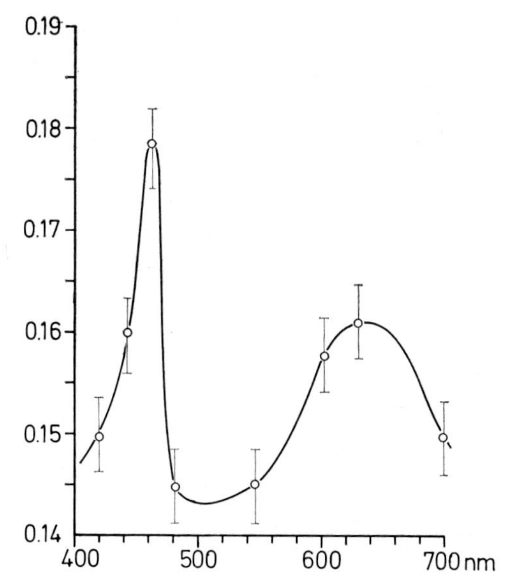

FIG. 32. Photokinetic action spectrum of *Euglena gracilis* (after Wolken and Shin, 1958).

swimming at various wavelengths. As shown in Fig. 32, there is one sharp maximum in the blue at 460 nm and a second smaller but broader one in the red between 600 and 700 nm. Wolken and Shin have suggested the blue maximum indicates a carotenoid (β-carotene) as the photoreceptor, but it is more consistent with chlorophyll *b*.

(d) *Diatoms.* The photokinetic action spectrum of the diatom *Nitzschia communis*, measured by Nultsch (1971), shows a main maximum in the red between 660 and 700 nm and a second smaller one in the blue at 440 nm (Fig. 33), which both agree with the *in vivo* absorption maxima of chlorophyll *a*. The wavelength position of these maxima and the activity in the far red point to photosystem I. The minor effectiveness of light between 500 and 550 nm, which is absorbed by fucoxanthin, is interpreted to indicate a slight participation of photosystem II (Nultsch, 1973a). The action spectrum must therefore be regarded as a mixed type (see above).

(e) *Red algae*. The same is true in the red alga *Porphyridium cruentum*, the action spectrum of which is shown in Fig. 34. The main maximum (560 nm) is found in the absorption range of B-phycoerythrin. Red

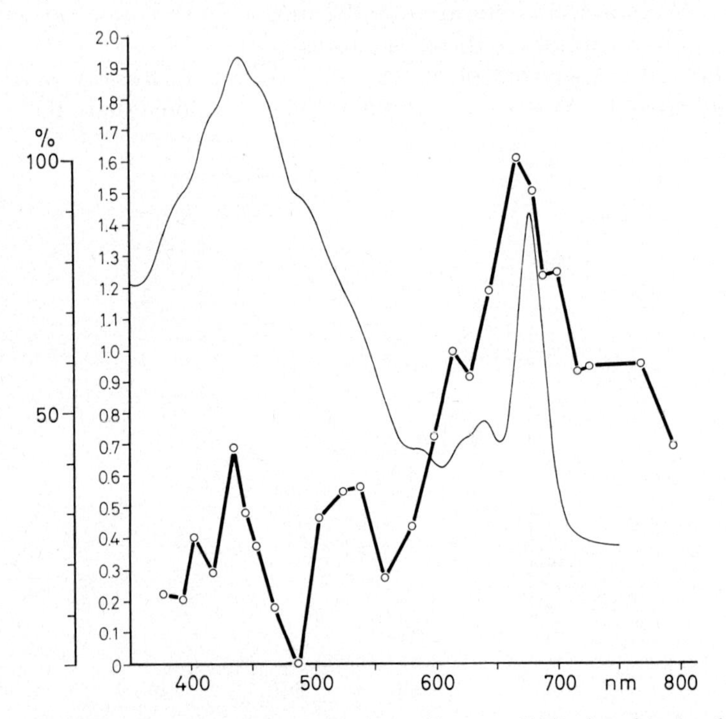

FIG. 33. Photokinetic action spectrum (circles and heavy solid line) and *in vivo* absorption spectrum (fine line) of *Nitzschia communis* (after Nultsch, 1971).

light absorbed by chlorophyll *a* is more effective than blue light absorbed by the Soret band, as is the case in *Nitzschia communis*. The two maxima between 600 and 650 nm cannot easily be interpreted if compared with the *in vivo* absorption spectrum, but they correspond fairly well with the absorption maxima of R- and allo-phycocyanin.

D. *Effect of inhibitors*

Although the photokinetic action spectra of the various groups investigated display considerable differences in detail, they have one feature in common: the photokinetically active light is absorbed by photosynthetic pigments. Since movement needs energy, the photokinetic effect could consist in an additional ATP supply from photosynthetic phosphorylation to the locomotor apparatus, and the action spectra could be determined by whether the energy source is cyclic photophosphorylation driven by photosystem I or a phosphorylation type in which both the photosystems are active. In the latter case differences in the accessory pigment content are of importance.

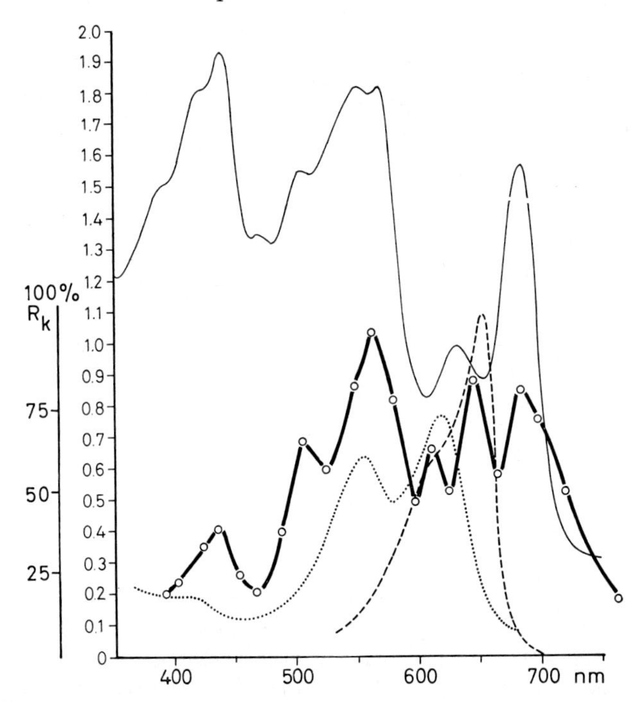

FIG. 34. Photokinetic action spectrum (circles and heavy solid line) and *in vivo* absorption spectrum (fine line) of *Porphyridium cruentum*. Absorption spectra of R-phycocyanin (dotted line) and allophycocyanin (dashed line) (after Nultsch and Dillenburger, from Nultsch, 1974a).

If the above views are correct, photokinesis should be inhibited by uncouplers and should display differences in sensitivity to various photosynthesis inhibitors. The effect of uncouplers and photosynthesis inhibitors on movement in light and in darkness was therefore investigated, especially in the "pure" types *Phormidium* and *Anabaena*. The effect of externally added redox-systems on movement was also studied.

(a) *Uncouplers*. The effect of uncouplers on the blue-green alga *Phormidium uncinatum* was studied by Nultsch (1967, 1969, 1974a). In general movement is inhibited by uncouplers such as DNP (=dinitrophenol), desaspidin and CCCP (=carbonylcyanide-m-chlorophenyl-hydrazone) in light and in darkness at concentrations comparable to those inhibiting photosynthetic phosphorylation and oxidative phosphorylation respectively. Thus it seems that movement in the dark is coupled with oxidative phosphorylation and that the photokinetic acceleration of movement is the result of an additional ATP supply from photosynthetic phosphorylation. The inhibition of motility by several uncouplers was also observed by Diehn and Tollin (1967) in *Euglena gracilis* and by Stavis and Hirschberg (1973) in *Chlamydomonas reinhardtii.*

(b) *Photosynthesis inhibitors.* The effect of some photosynthesis inhibitors, such as phenylurethane, hydroxylamine, *o*-phenanthroline and salicylaldoxime on the motility of some *Phormidium* species was investigated

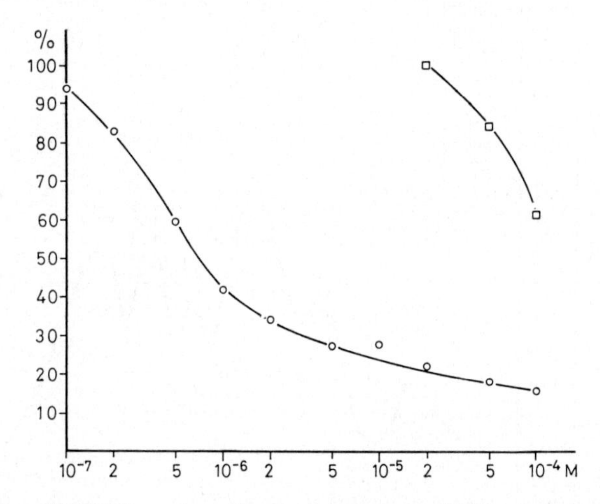

FIG. 35. Effect of DCMU on photokinesis of *Anabaena variabilis* (circles) and *Phormidium ambiguum* (squares) in white light (1000 lux). Abscissa: molar concentration; ordinate: speed of movement as percentage of that in controls (after Nultsch and Hellmann, 1972).

by Nultsch and Jeeji-Bai (1966). Although these substances inhibit photokinesis, most of them are not specific, since they inhibit movement in the dark even more than movement in the light, with the exception of DCMU (=3-(3·4-dichlorophenyl)-1·1-dimethylurea). Later investigations were therefore focussed on this and other specific inhibitors (Nultsch, 1969, 1973a, 1974a; Nultsch and Hellmann, 1972).

If we compare the effect of DCMU on the two "pure" photokinesis types (Fig. 35), we find *Anabaena variabilis* much more sensitive to this substance than *Phormidium ambiguum* and others. Since DCMU at low concentrations (10^{-6} M) specifically inhibits non-cyclic electron transport, while above 5×10^{-6} M the cyclic electron transport is also impaired (Gingras *et al.*, 1963; van Rensen, 1969), these experiments give further evidence that in the *Phormidium* type the acceleration of movement by light is the result of an ATP supply from cyclic photophosphorylation which is driven by photosystem I, whereas in *Anabaena* the energy is supplied by a photophosphorylation process in which photosystem II is also active. Diehn and Tollin (1967) and Stavis and Hirschberg (1973) found that DCMU does not impair the motility of *Euglena* and *Chlamydomonas*. They concluded that in these organisms the energy for the locomotor apparatus is supplied solely by oxidative phosphorylation. If this is correct, it is difficult to explain the photokinetic effect in *Euglena* demonstrated by Wolken and Shin or its action spectrum (see above).

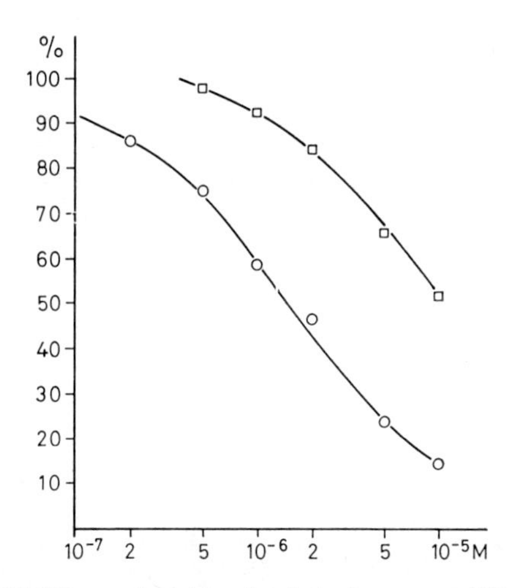

FIG. 36. Effect of DBMIB on photokinesis of *Anabaena variabilis* (circles) and *Phormidium ambiguum* (squares) in white light (1000 lux). Abscissa: molar concentrations; ordinate: speed of movement as percentage of that in controls (after Nultsch and Hellmann, 1972).

Photokinesis of *Anabaena variabilis* is more sensitive to the plastoquinone antagonist (Trebst *et al.*, 1970) DBMIB (=2·5-dibromo-3-methyl-6-isopropyl-p-benzoquinone) than *Phormidium ambiguum* (Fig. 36). This is understandable if we suppose that the cyclic electron flow at least partly bypasses plastoquinone (Böhme *et al.*, 1971).

The effect of DSPD (=disalicylidenepropane-diamin-1·3) on photokinesis was investigated by Nultsch (1974a). This substance inhibits the reduction of ferredoxin and, hence, NADP-reduction and CO_2-fixation (Trebst and Burba, 1967). Cyclic electron transport is more sensitive to it than non-cyclic. As shown in Fig. 37, *Anabaena variabilis* is a little more sensitive to this substance than *Phormidium ambiguum*. This would fit well into the above mentioned scheme. However, it seems doubtful whether the effect of DSPD is as specific as Trebst and Burba

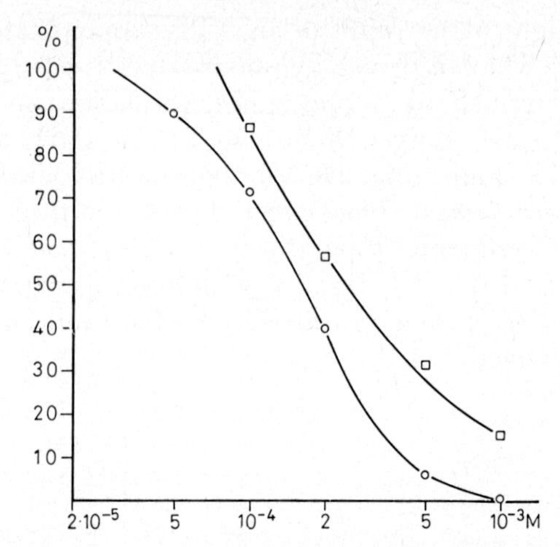

FIG. 37. Effect of DSPD on photokinesis of *Anabaena variabilis* (circles) and *Phormidium ambiguum* (squares) in white light (1000 lux). Abscissa: molar concentration; ordinate: speed of movement as percentage of that in controls (after Nultsch, 1974a).

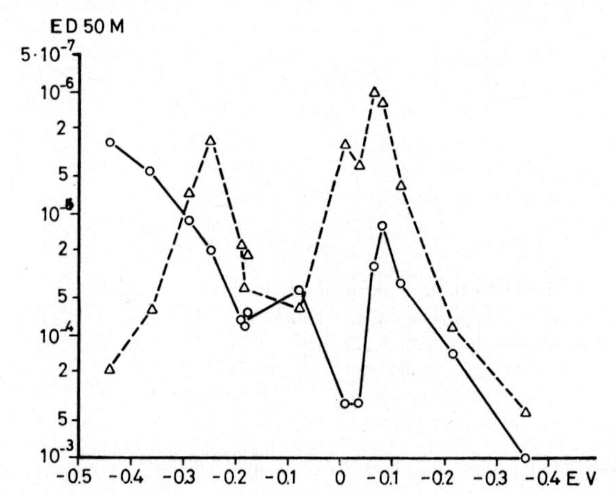

FIG. 38. Effect of redox systems on photokinesis (circles, solid line) and dark movement (triangles, dashed line) of *Phormidium uncinatum*. From left to right: methylviologen, benzylviologen, safranine T, phenosafranine, janus green, riboflavin, phthiocol, triphenyltetrazolium chloride, methylene blue, toluidine blue, thionin, phenazine methosulphate, toluylene blue, dichlorophenol-indophenol, ferricyanide. Abscissa: mid-point redox potential (E_0') in volts; ordinate: molar ED 50 (after Nultsch, 1968).

suggest, since movement in the dark is also impaired by DSPD and one should not expect that a ferredoxin inhibitor would play an important role in oxidative phosphorylation.

(c) *Redox systems*. Since artificial redox substances can serve as electron donors or electron acceptors in photosynthetic (Witt *et al.*, 1965, 1966) as well as in respiratory electron transport (Judah and Williams-Ashman, 1951; Racker, 1961), the effect of redox substances on the speed of movement in light and darkness of *Phormidium uncinatum* was investigated (Nultsch, 1968, 1974a). With the aid of dose-response-curves, the ED_{50}-values (the molar concentrations which inhibit the motility to 50% of that of the control) were evaluated and plotted against the midpoint potential of the redox systems. The curves obtained are shown in Fig. 38. Maximum inhibition of photokinesis occurs with viologens below −0·4 V, while a second smaller maximum lies between +0·06 and +0·08 V, i.e. the mid-point potentials of phenazinemethosulphate and thionine. Thus all redox systems which can trap electrons from the photosynthetic electron transport chain inhibit photokinesis but those which can drain them from cyclic electron transport are more effective. Movement in the dark is also impaired by all the redox substances investigated, since most of them can trap electrons from the respiratory chain. Two maxima were found, one at −0·29 V, i.e. the midpoint potential of safranin T, and the other at +0·1 V. These maxima are very close to the midpoint potentials of NAD^+ and NAD-cytochrome-reductase respectively.

E. *Effect of ATP*

If photokinesis is due to an additional ATP supply from photosynthetic phosphorylation, it should be possible to simulate the effect of light by adding ATP. However, all attempts to increase the speed of purple bacteria, blue-green algae and diatoms by externally added ATP were unsuccessful (Clayton, 1958; Nultsch, 1970). In these experiments the speed was measured immediately after ATP application. Throm (1968), however, who carried out long term experiments, observed that 10^{-3} M ATP accelerated the movement of *Rhodospirillum rubrum*, and that the longer it could act on the organisms the greater the effect. It was possible to saturate the locomotor apparatus so that light had no effect on the speed of movement. Hence, light and ATP are equivalent in accelerating movement (Nultsch and Throm, 1968). ATP cannot be replaced by GTP, CTP, ITP or other energy rich compounds (Nultsch and Throm, unpublished).

F. *Conclusions*

The experiments reported above indicate that photokinesis is due to an additional ATP supply from photosynthetic phosphorylation to the locomotor apparatus, and that cyclic photophosphorylation, a phosphorylation coupled with the non-cyclic electron transport or both can serve as energy sources. Movement in the dark, if it occurs, is maintained by ATP supplied from oxidative phosphorylation; anaerobic phosphorylation in glycolysis has not been found to play a role.

In 1882 Engelmann concluded that in micro-organisms a pool of a certain substance may be produced by light which is necessary for movement and is consumed in the dark. It is now established that the substance predicted by Engelmann is ATP.

V. Summary

The occurrence of three different types of reaction to light, having effects on movement—photo-topotaxis, photo-phobotaxis and photokinesis—has been clearly demonstrated in blue-green algae. As shown in Table 1

TABLE 1. Threshold values, optima and maxima in lux for light effects on movement of *Phormidium uncinatum*.

	Threshold[a]	Optimum	Maximum[b]
Topotaxis	2	200	10,000
Phobotaxis	0·1	5,000	50,000
Photokinesis	0·05	2,000	40,000

[a] These figures are for the zero threshold, i.e. the intensity needed to produce a response in previously dark-adapted cells.
[b] The value at which the response becomes negative.

and Fig. 39 for *Phormidium uncinatum*, they display different threshold values, optima and maxima, differ in their spectral sensitivity and display different sensitivities to inhibitors. Thus they are mediated by different photoreceptors and governed by different mechanisms. The mechanism of photo-topotaxis is still an unsolved problem, but there is strong evidence that photokinesis is the result of an additional ATP supply from photophosphorylation, and that the phobic responses are caused by sudden changes in the electron flow in which photosystems I and II have a role differing from species to species. In purple bacteria, in which an action spectrum for photo-topotaxis has not yet been measured, photo-phobotaxis and photokinesis are also mediated by the photosynthetic apparatus. Since purple bacteria lack a second light reaction in photosynthesis, the action spectra of photokinesis and photo-phobotaxis display no fundamental differences; some differences reported in the carotenoid region may be due to variations in the carotenoid content of the cells due to differing culture conditions (Clayton, 1959). However, in these organisms also the two processes seem to be coupled to photosynthesis in different ways: photo-phobotaxis to photosynthetic electron transport and photokinesis to photophosphorylation.

In eukaryotic organisms photokinesis has rarely been investigated. Action spectra obtained for *Euglena*, the diatom *Nitzschia communis* and the red alga *Porphyridium cruentum* indicate also a relationship to photosynthesis, and it seems possible that in all organisms the positive photokinetic effect is the result of an energy supply from photophosphorylation

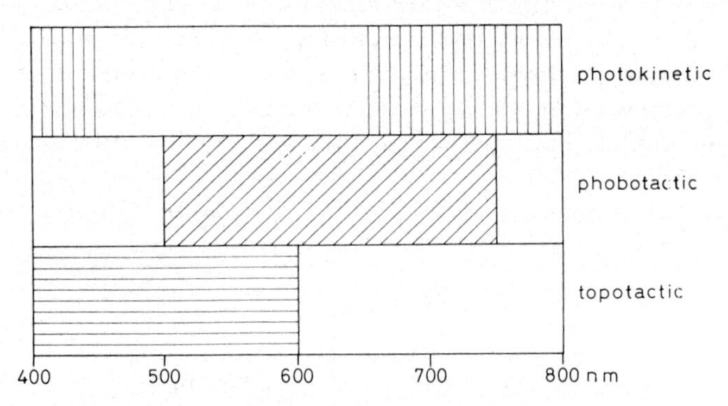

FIG. 39. Effective wavelengths for motility responses in *Phormidium uncinatum*.

to the locomotor apparatus. Two types of photo-phobotaxis must be distinguished in eukaryotes: the short wavelength type represented mainly by the flagellates and the photosynthetic type demonstrated with *Navicula peregrina* and *Porphyridium cruentum*. It is possible but not yet proved that in the latter the photophobic responses are also the result of sudden changes in electron transport. The photoreceptor of the short wavelength type remains unknown, although the action spectra suggest carotenoids or flavins. Since similar action spectra have been obtained for photophobic and photo-topotactic reactions in flagellates, in these organisms both reactions are probably mediated by the same photoreceptor. This could also be possible in the diatom *Nitzschia communis*, but is obviously not true in *Porphyridium cruentum*, in which the two action spectra differ considerably. Of great interest is the topotactic action spectrum of the blue-green alga *Anabaena variabilis* indicating correlations between photo-topotaxis and photosynthesis.

In some organisms, especially in flagellates, photo-topotactic orientation is brought about by a steering act, the details of which depend upon the morphology of the cells and the movement mechanism. In *Euglena* photo-orientation is supposed to be the result of numerous successive phobic responses which would explain the similarity of the topotactic and phobotactic action spectra. Even in the blue-green alga *Anabaena variabilis* phototopotaxis is the result of an active orientation. In other organisms (such as blue-green algae of the genus *Phormidium* and diatoms) no steering act exists, but in cells oriented more or less parallel to the light beam movements towards the light source last longer than those away from it, due to the front half receiving more quanta per time unit than the rear. Phobotactic reactions occur when the front receives less quanta per time unit than before.

There are many unanswered questions in the field of photomotion, and some of the suggested mechanisms are only working hypotheses. The sensory transduction from the receptor to the effector is an unsolved problem in nearly all the systems so far investigated. Since there are several different mechanisms in phototaxis and in movement, we cannot expect a universal mode of sensory transduction. However, several promising models have been proposed and further progress may be anticipated.

The author is indebted to Prof. Dr W. Haupt for valuable criticism.

References

ADERHOLD, R. (1888). *Jena. Z. Naturw.* **22,** (N.F. 15), 310–342.

ADLER, J. and DAHL, M. M. (1967). *J. Gen. Microbiol.* **46,** 161–173.

AMESZ, J. (1973). *Fortschr. Bot.* **35,** 89–102.

ASCOLI, C., BARBI, M., FREDIANI, C., PETRACCHI, D. and TRIMARCO, D. (1971). *Atti del I Congresso di Cibernetica, Vol. I,* p. 296, Litofelici, Pisa.

BATRA, P. and TOLLIN, G. (1964). *Biochim. Biophys. Acta* **79,** 371–378.

BENDIX, S. (1960). *In* "Comparative Biochemistry of Photoreactive Systems" (M. B. Allen, ed.), pp. 107–127, Academic Press Inc., New York.

BÖHME, H., REIMER, S. and TREBST, A. (1971). *Z. Naturforschung* **26b,** 341–352.

BOLTE, E. (1920). *Jb. wiss. Bot.* **59,** 287–324.

BOVEE, E. C. and JAHN, TH. L. (1972). *J. theor. Biol.* **35,** 259–276.

BRINKMANN, K. (1966). *Planta.* **70,** 344–389.

BRUCE, V. G. (1970). *J. Protozool.* **17,** 328–334.

BRUCE, V. G. and PITTENDRIGH, C. S. (1956). *Proc. Nat. Acad. Sci., U.S.A.,* **42,** 676–682.

BUDER, J. (1915). *Jb. wiss. Bot.* **56,** 529–584.

BUDER, J. (1917). *Jb. wiss. Bot.* **58,** 105–220.

BUDER, J. (1919). *Jb. wiss. Bot.* **58,** 525–628.

BÜNNING, E. and SCHNEIDERHÖHN, G. (1956). *Arch. Mikrobiol.* **24,** 80–90.

CHECCUCCI, A. (1973). *In* "Primary Molecular Events in Photobiology" (A. Checcucci and R. A. Weale, eds.), pp. 217–244, Elsevier Scientific Publishing Company, Amsterdam.

CHECCUCCI, A., COLOMBETTI, G., DEL CARRATORE, G., FERRARA, R. and LENCI, F. (1974). *Photochem. Photobiol.* **19,** 223–226.

CLAYTON, R. K. (1953a). *Arch. Mikrobiol.* **19,** 107–124.

CLAYTON, R. K. (1953b). *Arch. Mikrobiol.* **19,** 125–140.
CLAYTON, R. K. (1953c). *Arch. Mikrobiol.* **19,** 141–164.
CLAYTON, R. K. (1957). *Arch. Mikrobiol.* **27,** 311–319.
CLAYTON, R. K. (1958). *Arch. Mikrobiol.* **29,** 189–212.
CLAYTON, R. K. (1959). *In* "Handbuch der Pflanzenphysiology" (W. Ruhland, ed.), Vol. 17/1, pp. 371–387, Springer, Berlin, Göttingen, Heidelberg.
CLAYTON, R. K. (1964). *In* "Photophysiology" (A. C. Giese, ed.), Vol. II, pp. 51–77, Academic Press, New York.
DAVENPORT, D. (1973). *In* "Behaviour of Microorganisms" (A. Pérez-Miravete, ed.), pp. 106–116, Plenum Press, London.
DAVENPORT, D., WRIGHT, C. A. and CAUSLEY, D. (1962). *Science* **135,** 1059–1060.
DAVENPORT, D., CULLER, G. J., GREAVES, J. O. B., FORWARD, R. and HAND, W. G. (1970). *Trans. Biomed. Eng.* I.E.E.E. BME-17, 230–237.
DIEHN, B. (1969a). *Biochim. Biophys. Acta* **177,** 136–143.
DIEHN, B. (1969b). *Nature* **221,** 366–367.
DIEHN, B. (1969c). *Exp. Cell Res.* **56,** 375–381.
DIEHN, B. (1970). *Photochem. Photobiol.* **11,** 407–418.
DIEHN, B. (1973). *Science* **181,** 1009–1015.
DIEHN, B. and TOLLIN, G. (1966). *Photochem. Photobiol.* **5,** 523–535.
DIEHN, B. and TOLLIN, G. (1967). *Arch. Biochem. Biophys.* **121,** 169–177.
DIEHN, B. and KINT, B. (1970). *Physiol. Chem. Phys.* **2,** 483–488.
DREWS, G. (1957). *Ber. dt. Bot. Ges.* **70,** 259–262.
DREWS, G. (1959). *Arch. Protistenk.* **104,** 389–430.
DUYSENS, L. N. M. (1952). *Thesis, Utrecht.*
ECKERT, R. (1972). *Science* **176,** 473–481.
EDMONDSON, D. E. and TOLLIN, G. (1971). *Biochemistry* **10,** 113–124.
ENGELMANN, TH.W. (1882). *Pflügers Arch. ges. Physiol.* **29,** 387–400.
ENGELMANN, TH.W. (1883). *Pflügers Arch. ges. Physiol.* **30,** 95–124.
ENGELMANN, TH.W. (1888). *Bot. Ztg.* **46,** 661–669.
FEINLEIB, M. E. H. and CURRY, G. M. (1967). *Physiol. Plant.* **20,** 1083–1095.
FEINLEIB, M. E. H. and CURRY, G. M. (1971a). *In* "Handbook of Sensory Physiology" (W. R. Lowenstein, ed.), Vol. I, 366–395, Springer, Berlin, Heidelberg, New York.
FEINLEIB, M. E. H. and CURRY, G. M. (1971b). *Physiol. Plant.* **25,** 346–357.
FORWARD, R. B. Jr. (1970). *Planta* **92,** 248–258.
FORWARD, R. B. Jr. (1973). *Planta* **111,** 167–178.
FORWARD, R. B. Jr. and DAVENPORT, D. (1968). *Science* **161,** 1028–1029.
FRANCIS, D. W. (1964). *J. Cell Comp. Physiol.* **64,** 131–138.
FROEHLICH, O. and DIEHN, B. (1974). *Nature* **248,** 802–804.
GERISCH, G. (1959). *Arch. Protistenk.* **104,** 292–358.
GIBBONS, B. H. and GIBBONS, I. R. (1972). *J. Cell Biol.* **54,** 75–97.
GINGRAS, G., LEMASSON, C. and FORK, D. C. (1963). *Biochim. Biophys. Acta* **69,** 438–440.
GÖSSEL, I. (1957). *Arch. Mikrobiol.* **27,** 288–305.
GRAY, J. J. (1955). *J. Exp. Biol.* **32,** 775–801.
HAEDER, D.-P. (1973). *Thesis*, Marburg.

HAEDER, D.-P. (1974). *Arch. Mikrobiol.* **96,** 255–266.
HAEDER, D.-P. and NULTSCH, W. (1973). *Photochem. Photobiol.* **18,** 311–317.
HALLDAL, P. (1957). *Nature* **179,** 215–216.
HALLDAL, P. (1958). *Physiol. Plant.* **11,** 118–153.
HALLDAL, P. (1961). *Physiol. Plant.* **14,** 133–139.
HALLDAL, P. (1962). *In* "Physiology and Biochemistry of Algae" (R. A. Lewin, ed.), pp. 583–593. Academic Press, New York.
HALLDAL, P. (1963). *Ber. dt. Bot. Ges.* **76,** 323–327.
HALLDAL, P. (1964): *In* "Biochem. and Physiol. of Protozoa" (S. H. Hutner, ed.), Vol. III, pp. 277–296, Academic Press, New York.
HAND, W. G., COLLARD, P. A. and DAVENPORT, D. (1965). *Biol. Bull.* **128,** 90–101.
HAND, W. G., FORWARD, R. and DAVENPORT, D. (1967). *Biol. Bull.* **133,** 150–165.
HAND, W. G. and DAVENPORT, D. (1970). *In* "Photobiology of Micro-organisms" (P. Halldal, ed.), pp. 253–282. Wiley, London.
HAND, W. G. and HAUPT, W. (1971). *J. Protozool.* **18,** 361–364.
HARDER, R. (1920). *Z. Bot.* **12,** 353–462.
HARTSHORNE, J. N. (1953). *New Phytologist* **52,** 292–297.
HAUPT, W. (1959). *In* "Handbuch der Pflanzenphysiologie" (W. Ruhland, ed.), 17/1, pp. 318–370. Springer, Berlin, Göttingen, Heidelberg.
HAUPT, W. (1965). *Ann. Rev. Plant Physiol.* **16,** 267–290.
HAUPT, W. (1966). *Int. Rev. Cytol.* **19,** 267–299.
HAXO, F. T. (1960). *In* "Handbuch der Pflanzenphysiologie" (W. Ruhland, ed.), V/2, pp. 349–363, Springer, Berlin.
HAXO, F. T. and NORRIS, P. S. (1953). *Biol. Bull.* **105,** 374.
HAXO, F. T. and CLENDENNING, K. A. (1953). *Biol. Bull.* **105,** 103–114.
HEIDINGSFELD, J. (1943). *Thesis*, Breslau.
HESS, B. and OESTERHELT, D. (1974). *In* "Dynamics of Energy Transducing Membranes" (Ernster, Estabrook and Slater, eds.). Elsevier Publishing Comp., Amsterdam.
HILDEBRAND, E. and DENCHER, N. (1974). *Ber. dt. Bot. Ges.* **87,** 93–99.
HOLMES, S. J. (1903). *Biol. Bull.* **4,** 319–326.
HUTH, K. (1970a). *Z. Pflanzenphysiol.* **62,** 436–450.
HUTH, K. (1970b). *Z. Pflanzenphysiol.* **63,** 344–351.
JENNINGS, H. S. (1906). Behaviour of the lower organisms. Columbia University Press.
JUDAH, J. D. and WILLIAMS-ASHMAN, G. (1951). *Biochem. J.* **48,** 33–42.
KLEBS, G. (1885). *Biol. Zbl.* **5,** 353–367.
KRINSKY, N. I. and GOLDSMITH, T. H. (1960). *Arch. Biochem. Biophys.* **91,** 271–279.
KUNG, C. (1971). *Z. vergl. Physiol.* **71,** 142–164.
LINDES, D. A., DIEHN, B. and TOLLIN, G. (1965). *Rev. scient. Instrum.* **36,** 1721–1725.
LINKS, J. (1955). *Thesis*, Leiden.
LUNTZ, A. (1931a). *Z. vergl. Physiol.* **14,** 68–92.
LUNTZ, A. (1931b). *Z. vergl. Physiol.* **15,** 652–678.
LUNTZ, A. (1932). *Z. vergl. Physiol.* **16,** 204–217.

MAINX, F. (1929). *Arch. Protistenk.* **68,** 105–176.
MANTEN, A. (1948). *Thesis*, Utrecht.
MARBACH, I. and MAYER, A. M. (1970). *Phycologia* **9,** 255–260.
MASSART, J. (1902). *Biol. Zbl.* **22,** 9–23, 41–52, 65–79.
MAST, O. S. (1911). Light and the behaviour of organisms. Wiley, New York.
MAST, O. S. (1926). *Z. vergl. Physiol.* **4,** 637–685.
METZNER, P. (1929). *Z. Bot.* **22,** 225–265.
MOLISCH, H. (1907). Die Purpurbakterien nach neuen Untersuchungen. Gustav Fischer, Jena.
NAGEL, W. A. (1901). *Bot. Ztg.* **59,** 287–299.
NEUSCHELER, W. (1967a). *Z. Pflanzenphysiol.* **57,** 46–59.
NEUSCHELER, W. (1967b). *Z. Pflanzenphysiol.* **57,** 151–171.
NULTSCH, W. (1956). *Arch. Protistenk,* **101,** 1–68.
NULTSCH, W. (1961). *Planta* **56,** 632–647.
NULTSCH, W. (1962a). *Planta* **57,** 613–623.
NULTSCH, W. (1962b). *Planta* **58,** 647–663.
NULTSCH, W. (1962c). *Ber. dt. Bot. Ges.* **75,** 443–453.
NULTSCH, W. (1965). *Photochem. Photobiol.* **4,** 613–619.
NULTSCH, W. (1966). *Arch. Mikrobiol.* **55,** 187–199.
NULTSCH, W. (1967). *Z. Pflanzenphysiol.* **56,** 1–11.
NULTSCH, W. (1968). *Arch. Mikrobiol.* **63,** 292–320.
NULTSCH, W. (1969). *Photochem. Photobiol.* **10,** 119–123.
NULTSCH, W. (1970). *In* "Photobiology of Microorganisms" (P. Halldal, ed.), pp. 213–249. Wiley, London.
NULTSCH, W. (1971). *Photochem. Photobiol.* **14,** 705–712.
NULTSCH, W. (1973a). *In* "Primary Molecular Events in Photobiology" (A. Checcucci and R. A. Weale, eds.), pp. 245–273, Elsevier Scientific Publishing Company, Amsterdam.
NULTSCH, W. (1973b). *In* "Behaviour of Microorganisms" (A. Pérez-Miravete, ed.), Plenum Press, London.
NULTSCH, W. (1974a). Abhandl. Marburger Gelehrten Gesellschaft, Bd. 2, 138–213, Fink Verlag, München.
NULTSCH, W. (1974b). *In* "Algal Physiology and Biochemistry" (W. D. P. Stewart, ed.), 864–893, Blackwell Scientific Publications, Oxford.
NULTSCH, W. and HAEDER, D.-P. (1970). *Ber. dt. Bot. Ges.* **83,** 185–192.
NULTSCH, W. and HAEDER, D.-P. (1974). *Ber. dt. Bot. Ges.* **87,** 83–92.
NULTSCH, W. and HELLMANN, W. (1972). *Arch. Mikrobiol.* **82,** 76–90.
NULTSCH, W. and HOFF, E. (1973). *Arch. Protistenk.* **115,** 336–352.
NULTSCH, W. and JEEJI-BAI (1966). *Z. Pflanzenphysiol.* **54,** 84–98.
NULTSCH, W. and RICHTER, G. (1963). *Arch. Mikrobiol.* **47,** 207–213.
NULTSCH, W. and THROM, G. (1968). *Nature* **218,** 697–699.
NULTSCH, W. and THROM, G. (1975). *Arch. Mikrobiol.* **103,** 175–179.
NULTSCH, W., THROM, G. and v. RIMSCHA, I. (1971). *Arch. Mikrobiol.* **80,** 351–369.
NULTSCH, W. and WENDEROTH, K. (1973). *Arch. Mikrobiol.* **90,** 47–58.
OESTERHETT, O. and STOCKENINS, W. (1973). *Proc. Nat. Acad. Sci. U.S.A.* **70,** 2853–2857.
OJAKIAN, G. K. and KATZ, D. F. (1973). *Exp. Cell Res.* **81,** 487–491.

Oltmanns, F. (1917). *Z. Bot.* **9,** 257–338.
Pfeffer, W. (1904). *Pflanzenphysiologie*, 2. Aufl.
Philipps, D. M. (1972). *J. Cell Biol.* **53,** 561–573.
Poff, K. L. and Butler, W. L. (1974). *Nature* **248,** 799–801.
Poff, K. L., Butler, W. L. and Loomis, W. F. (1973). *Proc. Nat. Acad. Sci., U.S.A.* **70,** 813–816.
Pohl, R. (1948). *Z. Naturf.* **3b,** 367–374.
Pringsheim, E. G. (1968). *Arch. Mikrobiol.* **61,** 169–180.
Racker, E. (1961). *Advanc. Enzymol.* **23,** 323–399.
Van Rensen, J. J. S. (1969). *In* "Progress in photosynthesis research" (H. Metzner, ed.), Vol. III, pp. 1769–1776, Tübingen.
Rikmenspoel, R. and Van Herpen, G. (1957). *Phys. in Med. and Biol.* **2,** 54–63.
Ringo, D. L. (1967). *J. Cell Biol.* **33,** 543–571.
Robertson, J. A. (1972). *Arch. Mikrobiol.* **85,** 259–266.
Rothert, W. (1901). *Flora, Jena* **88,** 371–421.
Schilde, C. (1966). *Planta* **71,** 184–188.
Schlegel, H. G. (1956). *Arch. Protistenk.* **101,** 69–97.
Schneider, W. R. and Doetsch, R. N. (1974). *Appl. Microbiol.* **27,** 283–284.
Schrammeck, J. (1934). *Beitr. Biol. Pfl.* **22,** 315–379.
Stahl, E. (1880). *Bot. Z.* **38,** 297–413.
Stanier, R. Y., Kunisawa, R., Mandel, M. and Cohen-Bazire, G. (1971). *Bact. Rev.* **35,** 171–205.
Stavis, R. L. (1974). *Proc. Nat. Acad. Sci., U.S.A.* **71,** 1824–1827.
Stavis, R. L. and Hirschberg, R. (1973). *J. Cell Biol.* **59,** 367–377.
Strasburger, E. (1878). 'Wirkung des Lichtes und der Wärme auf Schwärmsporen.' Jena.
Thomas, J. B. (1950). *Biochim. Biophys. Acta* **5,** 186–196.
Thomas, J. B. and Nijenhuis, L. E. (1950). *Biochim. Biophys. Acta* **6,** 317–324.
Throm, G. (1968). *Arch. Protistenk.* **110,** 313–371.
Throm, G. (1970). *Z. Pflanzenphysiol.* **63,** 162–180.
Throm, G. (1971). *Z. Pflanzenphysiol.* **65,** 389–403.
Tollin, G. (1969). *In* "Current Topics in Bioenergetics" (D. R. Sanadie, ed.), Vol. III, 417–446, Academic Press, New York.
Trebst, A. and Burba, M. (1967). *Z. Pflanzenphysiol.* **57,** 419–433.
Trebst, A. and Harth, E. (1970). *Z. Naturfosch.* **25b,** 1157–1159.
Treviranus, L. G. (1817). Vermischte Schriften anatomischen und physiologischen Inhalts **2,** 71–92.
Uffen, R. L., Sybesma, C. and Wolfe, R. S. (1971). *J. Bacteriol.* **108,** 1348–1356.
Vávra, J. (1956). *Arch. Mikrobiol.* **25,** 223–225.
Wenderoth, K. (1975). *Thesis*, Marburg.
Witt, T. H., Rumberg, B., Schmidt-Mende, P., Sigel, U., Skerra, B., Vater, J. and Weikard, J. (1965). *Angew. Chem.* **77,** 821–876.
Witt, T. H., Skerra, B. and Vater, J. (1966). *In* "Currents in photosynthesis" (J. B. Thomas and J. C. Goedheer, eds.), pp. 273–292, A. D. Donker, Rotterdam.
Wolken, J. J. and Shin, E. (1958). *J. Protozool.* **5,** 39–46.

Chapter 3

Chemotaxis in bacteria

JULIUS ADLER

Departments of Biochemistry and Genetics, University of Wisconsin, Madison, Wisconsin, U.S.A.

I. Introduction

Although chemotaxis by bacteria has been recognized since the end of the 19th century, thanks to the pioneering work of Engelmann, Pfeffer, and other biologists (Weibull, 1960), the mechanisms involved are still almost entirely unknown. How do bacteria detect the attractants? How is this sensed information translated into action—that is, how are the flagella directed?

To learn about the detection mechanism that bacteria use in chemotaxis, it is important first to know *what* is being detected. One possibility is that the attractants themselves are detected. In that case, extensive metabolism of the attractants would not be necessary for chemotaxis. There is another possibility—that the attractants themselves are not detected but, instead, some metabolite of the attractants (for example, the pyruvate inside the cell) or the energy produced from the attractants, perhaps in the form of adenosine triphosphate. In these cases, metabolism of the attractants would be necessary for chemotaxis. The idea that bacteria sense the energy produced from the attractants at one time gained wide acceptance in the explanation of chemotaxis and phototaxis (Clayton, 1964; Links, 1955).

To try to determine which of these possible explanations is correct, experiments were carried out with *Escherichia coli* which exhibits chemotaxis toward various organic nutrients (Adler, 1966). The results show that extensive metabolism of the attractants is neither required nor sufficient for chemotaxis. Instead, the attractants themselves are detected. The systems that bacteria use to detect chemicals without metabolizing them are here called "chemoreceptors". Efforts to identify the chemoreceptors are described below.

II. A quantitative method for studying chemotaxis

Pfeffer (1884, 1888) demonstrated chemotaxis by pushing a capillary tube containing a solution of attractant into a suspension of motile bacteria on a slide and then observing microscopically that the bacteria accumulated first near the mouth of the capillary and later inside. A modification (Adler, 1969, 1973) of this method, which permits quantitative study of chemotaxis, is as follows. After incubation at 30°C for 60 minutes, the capillary is taken out of the bacterial suspension and the number of bacteria inside the capillary is measured by plating the contents of the capillary and counting colonies the next day. Reproducibility of the method is $\pm 15\%$. An attractant is tested over a range of concentrations usually between 10^{-8} M and 10^{-1} M in ten-fold intervals. From such an experiment one can construct a concentration-response curve and estimate a threshold concentration for accumulation inside the capillary. Some examples of attractants are galactose, glucose, ribose, aspartate, and serine, the sugars tested having the D and the amino acids the L configuration.

III. Evidence that the attractants themselves are detected

The following five approaches lead to the conclusion that chemotaxis is not a consequence of the metabolism of the attractants but that the attractants themselves are detected. More complete data was published by Adler (1969).

(a) *Some chemicals that are metabolized fail to attract bacteria.* These include galactonate, gluconate, glucuronate, glycerol, α-ketoglutarate, succinate, fumarate, malate, and pyruvate. This makes it clear that metabolism of a chemical and energy production from it are not sufficient to make a chemical an attractant.

(b) *Some chemicals that are nonmetabolizable attract bacteria.*

(i) *Mutant bacteria that have lost the ability to metabolize a chemical are attracted to it.* Mutants which lack three different enzymatic activities essential for the metabolism of galactose are attracted to galactose as well as are wild-type bacteria. These mutants are 99·5% or more blocked in their metabolism of galactose, relative to a wild-type strain (Adler, 1969). A similar result for taxis to glucose was obtained with a mutant defective in its ability to metabolize glucose.

(ii) *Some nonmetabolizable analogs of metabolizable chemicals attract bacteria.* D-Fucose (6-deoxy-D-galactose) is a galactose analog that is not metabolized by *E. coli* (Buttin, 1963; Adler, 1969). Nevertheless the bacteria are attracted to it very well, although its threshold concentration for chemotaxis is higher than that of D-galactose, as might be expected for an analog. The D-fucose had been purified to remove metabolizable impurities such as galactose or glucose. Three glucose analogs which were known to be non-metabolizable are also attractants: 2-deoxyglucose, α-methyl glucoside, and L-sorbose. The three analogs were purified before use in order to remove glucose or other contaminants.

(c) *Chemicals attract bacteria even in the presence of a metabolizable chemical.* If bacteria detect metabolites of an attractant, or energy produced from it, then the addition of a metabolizable chemical should stop chemotaxis by flooding the cells with metabolites and energy. This was not found to be the case for either metabolizable or nonmetabolizable attractants (Adler, 1969).

(d) *Attractants that are closely related in structure compete with each other but not with structurally unrelated compounds.* This finding supports the conclusion that it is the attractants themselves that are detected, and that there exists a variety of specific receptors. In these "competition experiments" one attractant, usually at 0·01 M, is put into the capillary tube, and another attractant at a concentration of 0·01 M is put into both the capillary and the bacterial suspension. If the two attractants use the same chemoreceptor, the response should be inhibited; if they do not, the response should not be affected. Only two examples will be presented here; for further examples see Adler (1969), Adler *et al.* (1973) and Mesibov and Adler (1972). Chemotaxis toward fucose was completely inhibited by the presence of galactose, and in the reciprocal experiment there was nearly complete inhibition. This suggests that fucose and galactose use the same chemoreceptor (the "galactose receptor"). Glucose completely eliminated taxis toward galactose, but in the reciprocal experiment the inhibition was only about 60–70%, no matter how high the concentration of galactose. This suggests that the receptor which detects galactose also detects glucose but that in addition there is another receptor that detects glucose but not galactose—the "glucose receptor".

(e) *There are mutants which fail to carry out chemotaxis to certain attractants but are still able to metabolize them.* If there are chemoreceptors in bacteria and if they are specific, there should be mutants that are defective in their response to some attractants but not to others, because of a defect in a single receptor. Such mutants of *E. coli* have now been found (Adler *et al.*, 1973; Mesibov and Adler, 1972; Hazelbauer *et al.*, 1969). One mutant, defective in "serine receptor" activity, fails to be attracted to serine and shows much-reduced taxis toward glycine, alanine, and cysteine. These residual responses result from the "aspartate receptor". The mutant is attracted normally to aspartate and glutamate, and to galactose, glucose, and ribose. It oxidizes and takes up L-serine at the same rate as its parent. Another mutant, lacking "aspartate receptor" activity, shows no chemotaxis toward aspartate and glutamate, nearly normal taxis toward serine, alanine, glycine, and cysteine, and normal taxis toward galactose, glucose, and ribose. The rate of oxidation and uptake of aspartate is the same for the mutant and its parent. A third type of mutant, without the "galactose receptor", is not attracted to galactose and fucose and is attracted to glucose at a higher than normal threshold. This residual response to glucose results from the "glucose receptor". These mutants are attracted normally to fructose, ribose, serine, and aspartate. Metabolism of galactose is normal in these mutants. However, in some of the mutants there is a defect in the uptake of galactose, which will be discussed below.

The existence of these three non-chemotactic mutants argues for specific receptors and provides additional support for the idea that detection of the attractants is independent of their metabolism.

IV. How many chemoreceptors?

To determine how many kinds of chemoreceptors there are, three approaches are being used. The first is to ask whether a given attractant is still effective when another attractant is present—see "competition experiments" above. The second is to try to isolate mutants defective in individual receptor activities. A third is to study the inducibility of specific taxes—presumably the inducibility of specific receptors. For example, taxis toward galactose and fucose is inducible by galactose. The conclusion from results obtained so far (Adler, 1969; Adler *et al.*, 1973; Mesibov and Adler, 1972; Hazelbauer *et al.*, 1969) is that for positive chemotaxis there are at least the eleven chemoreceptors shown in Table 1. Oxygen is also known to be an attractant for *E. coli* (Adler, 1969) so there could be a receptor for it, but this has not been investigated

TABLE 1. Partial list of chemoreceptors for positive chemotaxis in *Escherichia coli.*

Receptors and attractants	Threshold Molarity[a]
N-*acetyl-glucosamine receptor*	
N-Acetyl-D-glucosamine	1×10^{-5}
Fructose receptor	
D-Fructose	1×10^{-5}
Galactose receptor	
D-Galactose	4×10^{-7}
D-Glucose	4×10^{-7}
D-Fucose	3×10^{-5}
Glucose receptor	
D-Glucose	1×10^{-5}
D-Mannose	1×10^{-5}
Maltose receptor	
Maltose	2×10^{-6}
Mannitol receptor	
D-Mannitol	7×10^{-6}
Ribose receptor	
D-Ribose	3×10^{-7}
Sorbitol receptor	
D-Sorbitol	1×10^{-5}
Trehalose receptor	
Trehalose	6×10^{-6}
Aspartate receptor	
L-Aspartate	6×10^{-8}
L-Glutamate	2×10^{-5}
Serine receptor	
L-Serine	4×10^{-7}
L-Cysteine	6×10^{-6}
L-Alanine	7×10^{-5}
Glycine	2×10^{-5}

[a] The threshold values are lower in mutants unable to take up or metabolize a chemical.

so far. A survey of other possible attractants is incomplete. It is conceivable that as well as having chemoreceptors some bacteria might have receptors specialized to detect light, gravity or temperature since all these stimuli are thought to elicit tactic responses in some bacteria (Weibull, 1960).

V. Negative chemotaxis

Bacteria are repelled by certain chemicals—negative chemotaxis. Tso and Adler (1974) have developed methods for studying this, and

TABLE 2. Partial list of chemoreceptors for negative chemotaxis in *Escherichia coli*.

Receptors and attractants	Threshold Molarity
Fatty Acid Receptor	
Acetate (C2)	3×10^{-4}
Propionate (C3)	2×10^{-4}
Butyrate, *iso*-butyrate (C4)	10^{-4}
Valerate, *iso*-valerate (C5)	10^{-4}
Caproate (C6)	10^{-4}
Heptanoate (C7)	6×10^{-3}
Alcohol Receptor	
Methanol	10^{-1}
Ethanol	10^{-3}
n-Propanol	4×10^{-3}
iso-Propanol	6×10^{-4}
iso-Butanol	10^{-3}
Hydrophobic Amino Acid Receptor	
Leucine	10^{-4}
Isoleucine	$1{\cdot}5 \times 10^{-4}$
Valine	$2{\cdot}5 \times 10^{-4}$
Tryptophan	10^{-3}
Phenylalanine	3×10^{-3}
Glutamine	3×10^{-3}
Histidine	5×10^{-3}
Indole Receptor	
Indole	10^{-6}
Skatole	10^{-6}
Aromatic Receptor	
Benzoate	10^{-4}
Salicylate	10^{-4}
H^+ Receptor	
Low pH	pH 6·5
OH^- Receptor	
High pH	pH 7·5

have identified a large number of chemicals that are repellents for *E. coli*. By means of competition experiments and negative chemotaxis mutants, the repellents have been grouped into seven classes (Table 2). Although most of the repellents are harmful substances, not all of them are, and various harmful substances are not repellents. Therefore the mechanism for detection of repellents does not depend on harm done, but rather it seems that there are chemoreceptors that detect the repellents themselves (Tso and Adler, 1974).

VI. What is the nature of the chemoreceptors?

One possibility is that the chemoreceptors are the first enzymes in the metabolism of the chemicals. This possibility has been excluded in the case of galactose, because mutants that lack galactokinase still respond perfectly well. Another possibility is that the chemoreceptors are the permeases and related components essential for transport of substances into the cell. To find out, mutants defective with respect to transport have been investigated from the standpoint of chemotaxis.

An *E. coli* mutant, 20SOK$^-$, that is defective in the uptake of galactose (Buttin, 1963; Rotman *et al.*, 1968) to the extent of a 99·5% block is attracted to galactose perfectly well (Adler, 1969). Thus transport is not required for chemotaxis, and the chemoreceptors therefore appear to be located somewhere on the "outside" of the cell. The galactose binding protein (Anraku, 1968), a component needed for the transport of galactose (Boos, 1969), is present in this mutant (Hazelbauer and Adler, 1971), so it must lack an additional component needed for galactose transport. We (Hazelbauer and Adler, 1971) have now found that the galactose binding protein is also needed for galactose taxis, and that this is the component of the galactose receptor that recognizes galactose, glucose, and a number of structurally related chemicals. The evidence is as follows:

1. The specificity of the galactose binding protein is the same as the specificity of the galactose receptor.
2. Most of the galactose taxis mutants are defective in the transport of galactose, as mentioned above, and these all have very low levels of galactose binding protein.
3. Galactose taxis can be eliminated by a mild osmotic shock which releases the galactose binding protein from the cells (Anraku, 1968).

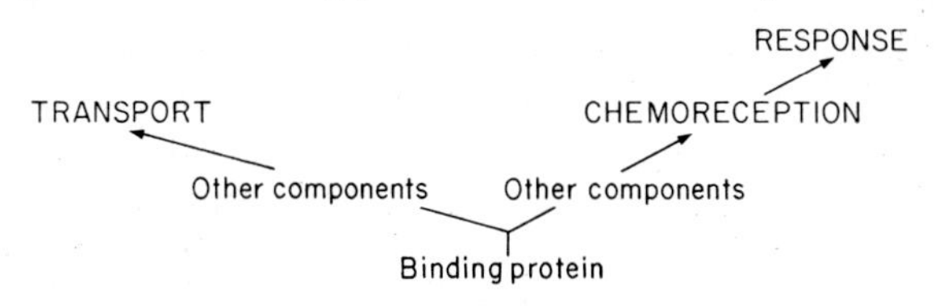

FIG. 1. Relationship between transport and chemoreception.

4. The galactose chemoreceptor is saturated at concentrations above 10^{-6} M, and so is the galactose binding protein (Mesibov *et al.*, 1973).

There must be one or more additional components of the galactose chemoreceptor, since at least one of the galactose taxis mutants has normal galactose binding protein (Hazelbauer and Adler, 1971). We are trying to find out what this additional component is. The role of the binding protein and the relationship between transport and taxis are summarized in Fig. 1.

We have also found binding proteins for maltose and ribose that are released by osmotic shock and appear to serve the corresponding chemoreceptors (Hazelbauer and Adler, 1971). For certain of the chemoreceptors, for example fructose, glucose, and mannitol, the binding proteins are the enzymes II of the phosphotransferase system (Adler *et al.*, 1973). These are not released by osmotic shock, being tightly bound to the cytoplasmic membrane.

VII. How do chemoreceptors work?

The mechanism of chemoreception in bacteria still remains unknown. Boos (1971) has shown that the galactose binding protein undergoes a configurational change when it binds galactose. Somehow this change may be "felt" by the additional components. It is the change in the fraction of binding protein occupied by attractant that is detected (Mesibov *et al.*, 1973). Then this information is transmitted to the flagella, either by a diffusible substance or by a configurational change of macromolecules perhaps in the cell membrane. The latter mechanism could involve something like a receptor potential or an action potential. The flagella then respond by changing their orientation in such a way to bring about a change in the frequency of cell "twiddling" (see below).

Bacteria in the absence of a gradient swim in a straight line for a while (a "run"), then they tumble—a "twiddle" or series of twiddles—then they again run in a new random direction (Berg and Brown, 1972) Macnab and Koshland (1972) and Berg and Brown (1972) have found that in a gradient of attractant the bacteria twiddle less frequently if they happen to swim toward the source of attractant and (of lesser significance) more frequently if away; in this way they migrate toward attractants.

We have isolated and reported on mutants which are generally non-chemotactic, i.e., they fail to carry out chemotaxis toward any of the attractants—amino acids, sugars, or oxygen—although the bacteria are perfectly motile (Armstrong *et al.*, 1967). Since it is unlikely that a single mutation would lead to a loss of all the kinds of chemoreceptors, these mutants are probably defective at some stage beyond the receptors, as shown diagrammatically in Fig. 2.

The defect could be in a transmitting system through which information from all the receptors is channeled to the flagella, or in the responding mechanism itself. Genetic analyses of the mutants have shown that three genes are involved (Armstrong and Adler, 1969a, b). Some of the mutants never tumble, while others always tumble. Further studies of these mutants may lead to an understanding of the way in which the chemoreceptors direct the flagella.

VIII. Conclusions

Extensive metabolism of chemicals is neither required, nor sufficient, for attraction of bacteria to the chemicals. Instead, the bacteria detect the attractants themselves. The systems that carry out this detection are called "chemoreceptors". There are mutants that fail to be attracted to one particular chemical or to a group of closely related chemicals but still metabolize these chemicals normally. These mutants are regarded as being defective in the activity of specific chemoreceptors. Data obtained so far indicate that there are at least eleven different chemoreceptors for positive chemotaxis in *Escherichia coli*. Negative chemotaxis

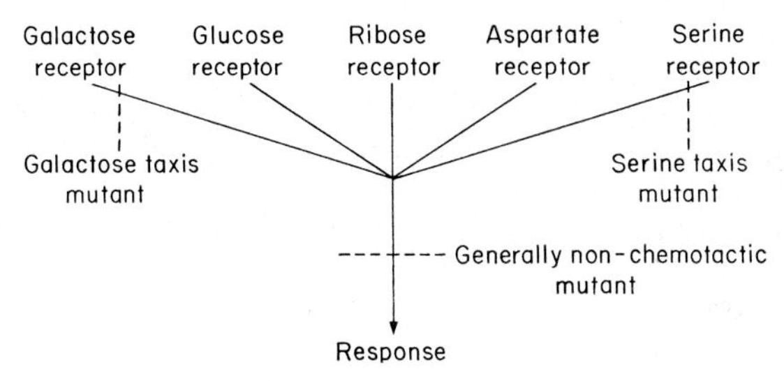

FIG. 2. The scheme for chemotaxis suggested in this article, showing location of defects in the various mutants.

has also been studied. A large number of repellents have been identified and classified into chemoreceptor groups. The chemoreceptors are not the enzymes that catalyze the metabolism of the attractants, nor are they certain parts of the permeases and related transport systems, and uptake itself is not required or sufficient for chemotaxis. In the case of the galactose receptor, the galactose binding protein is the component that recognizes the galactose.

Acknowledgments

The research discussed was supported by a grant from the U.S. National Institutes of Health. I thank Margaret Dahl for having carried out many of the experiments mentioned here, and my various students and post-doctoral fellows for major contributions. Part of this paper is reproduced from an article published in Science (Adler, 1969) by permission of the American Association for the Advancement of Science (copyright 1969).

References

ADLER, J. (1966). *Science* **153,** 708–716.
ADLER, J. (1969). *Science* **166,** 1588–1597.

Adler, J. (1973). *J. Gen. Microbiol.* **74,** 77–91.
Adler, J., Hazelbauer, G. L. and Dahl, M. M. (1973). *J. Bact.* **115,** 824–847.
Anraku, Y. (1968). *J. Biol. Chem.* **243,** 3123–3127.
Armstrong, J. B., Adler, J. and Dahl, M. (1967). *J. Bact.* **93,** 390–398.
Armstrong, J. B. and Adler, J. (1969a). *J. Bact.* **97,** 156–161.
Armstrong, J. B. and Adler, J. (1969b). *Genetics* **61,** 61–66.
Berg, H. C. and Brown, D. A. (1972). *Nature* **239,** 500–504.
Boos, W. (1969). *Europ. J. Biochem.* **10,** 66–73.
Boos, W. (1971). *Fed. Proc.* **30,** 1062.
Buttin, G. (1963). *J. Mol. Biol.* **7,** 164–182.
Clayton, R. K. (1964). *In* "Photophysiology" (A. C. Giese, ed.), Vol. 2, pp. 51–77. Academic Press, New York.
Hazelbauer, G. L., Mesibov, R. E. and Adler, J. (1969). *Proc. Nat. Acad. Sci. U.S.A.* **64,** 1300–1307.
Hazelbauer, G. L. and Adler, J. (1971). *Nature New Biol.* **230,** 101–104.
Links, J. (1955). Onderzoekingen Met Polytoma Uvella, Thesis, University of Leiden (in Dutch; summary in English).
Macnab, R. M. and Koshland, D. E., Jr. (1972). *Proc. Nat. Acad. Sci. U.S.A.* **69,** 2509–2512.
Mesibov, R., Ordal, G. W. and Adler, J. (1973). *J. Gen. Physiol.* **62,** 203–223.
Mesibov, R. E. and Adler, J. (1972). *J. Bact.* **112,** 315–326.
Pfeffer, W. (1884). *Untersuch. Botan. Inst. Tübingen* **1,** 363–482.
Pfeffer, W. (1888). *Untersuch. Botan. Inst. Tübingen* **2,** 582–661.
Rotman, B., Ganesan, A. K. and Guzman, R. (1968). *J. Mol. Biol.* **36,** 247–260.
Tso, W.-W. and Adler, J. (1974). *J. Bact.*, **118,** 560–576.
Weibull, C. (1960). *In* "The Bacteria" (I. C. Gunsalus and R. Y. Stanier, eds.), Vol. 1, pp. 153–205. Academic Press, New York.

Chapter 4

Chemotaxis in the cellular slime moulds

THEO M. KONIJN

Cell Biology and Morphogenesis Unit,
Zoologisch Laboratorium der Rijksuniversiteit te Leiden,
Leiden, Kaiserstraat 63, The Netherlands

I. Introduction

The cellular slime moulds (Acrasiales) are amoebae, common in forest soils, that undergo aggregation to form a "slug" or pseudoplasmodium just visible to the naked eye and subsequently differentiating into a fruiting body bearing spores (Fig. 1). The book by Bonner (1967) gives

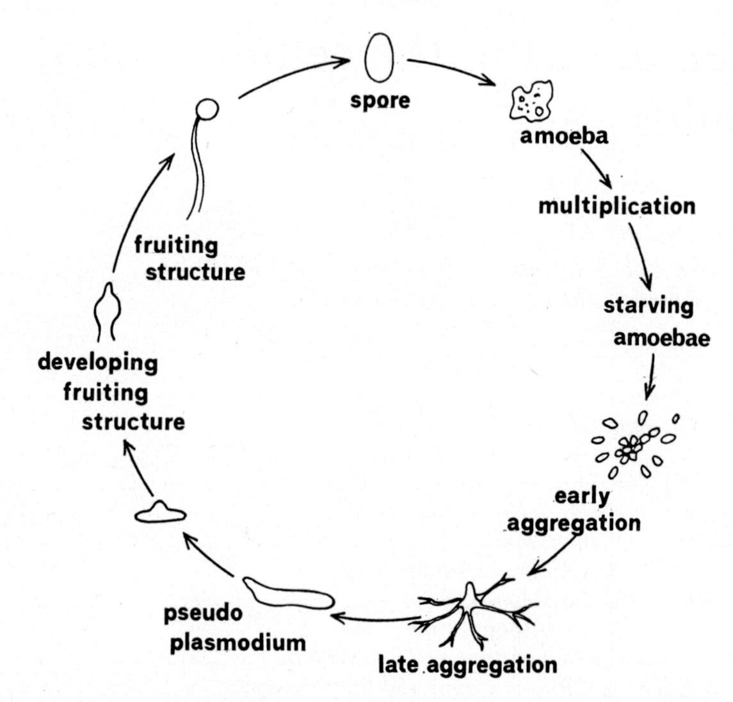

FIG. 1. Life cycle of *Dictyostelium discoideum*. Amoebae are about 10 μm in diameter and divide approximately every 3 hours. A pseudoplasmodium is commonly a few mm. long and 0·2–0·5 mm wide. It may contain 10 to 100,000 amoebae from which about a third differentiate into stalk cells and the rest into spores. The entire life-cycle, spore to spore, will take about two days.

an excellent account of the main species and their growth, morphogenetic movements and differentiation, and more recently the taxonomy of the Acrasiales (Acrasiomycetes) has been dealt with by Raper (1973).

In the context of the present volume the Acrasiales are of interest because they are the amoeboid organisms in which chemotaxis has been most intensively studied. Aggregation of amoebae to form a fruiting body first received attention, and when chemotaxis was found to be responsible, the attractant was termed "acrasin". Subsequently it was found that bacteria would attract amoebae, and this chemotaxis has also been studied. A negative chemotaxis to a gaseous factor seems to be involved in the orientation of fruiting structures, but this has received less attention.

Dictyostelium discoideum is the most intensively studied member of the Acrasiales and statements in the present chapter refer to this species, unless another species is named, or the Acrasiales as a whole are being discussed. "Vegetative amoebae" refers to amoebae that are growing, dividing and feeding on bacteria. "Pre-aggregative amoebae" refers to amoebae at the end of the vegetative phase, which have ceased feeding and are becoming sensitive to acrasin. Use of the term "acrasin" should be limited to the substances released during the aggregation process and which attract amoebae into the aggregate, and not applied to other substances which may attract amoebae. In *D. discoideum* and some other species it is clear that the acrasin is cyclic AMP; in another group of species cyclic AMP is not active in aggregation and another acrasin must exist.

II. Assays for chemotactic factors

The purification and identification of a chemotactic factor is dependent on the development of a reliable assay. An assay system can subsequently be used to measure concentrations of a chemotactic factor against standards of known concentration and thus to determine the amount present in various sources. It can be used to measure the effectiveness of various living sources under different environmental conditions and stages in the life-cycle. Conversely, using known concentrations of attractant, the sensitivity of responding cells from various species or developmental stages under a range of environmental conditions can be ascertained. Our conclusions about the effectiveness of attractants and the behaviour of responding cells are hence influenced by the nature and reliability of the assay. The development of chemotactic assays in the Acrasiales has been particularly difficult, and many have been devised. Apparent discrepancies between the results of different workers are often the consequence of employment of different assays. Hence it is important to take into account the potentialities and limitations of the various assays that have been used.

A. The discovery of acrasin and early assays

At the beginning of the century, Olive (1902) and Potts (1902) suggested that free amoebae respond to chemotactic signals emitted by aggregates of amoebae. Unsuccessful efforts were made to attract amoebae by malic acid and sugars, although Pfeffer had succeeded with those substances with spermatozoids of ferns.

The first experiment establishing that chemotaxis was involved in aggregation of amoebae was carried out by Runyon (1942). When he placed amoebae on both sides of a cellophane membrane, the amoebae on one side aggregated earlier, and when the cells on the other side reached the aggregative phase their streams coincided with those formed earlier. From this coincidence of aggregation patterns on both sides of the membrane Runyon concluded that a dialysable substance caused the chemotactic response of the amoebae.

Bonner (1947) tested the aggregative behaviour of amoebae on a glass surface under water. He placed the centre of an aggregate on one side of a coverslip and sensitive amoebae on the other, and showed that the latter moved around the edge of the coverslip to the opposite side, where they joined the centre. In another variation on his underwater experiments, two coverslips were placed side by side, one carrying near the edge a centre to which sensitive amoebae on the other coverslip responded. When the coverslips were separated, the amoebae joined the centre on the neighbouring coverslip when the distance was narrowed to 20–30 μm, thus showing that the chemical signal secreted by the centre was transmitted across the gap. Bonner's most convincing underwater technique used water flowing past an aggregate of *Dictyostelium*. Only the downstream amoebae moved to the centre; the upstream amoebae moved at random. The obvious conclusion was that the amoebae of the aggregate secreted a chemotactic substance that was washed downstream. In an attempt to quantify the underwater technique Bonner increased the distance between two coverslips both occupied by amoebae. Aggregating amoebae on one coverslip faced sensitive pre-aggregative amoebae on the opposite coverslip. The response of the younger amoebae was considered positive when they aggregated directly opposite a centre on the other coverslip. The older centre exerted a strong orientational effect over a distance of 500 μm and a weak effect when the distance between the two coverslips was increased to 800 μm.

After Runyon and Bonner demonstrated convincingly that chemotaxis controls cell aggregation in the cellular slime moulds, many attempts were made to identify the chemotactic factor, termed acrasin by Bonner, and several methods of isolating, purifying and finally identifying attracting compounds were developed.

B. *The sandwich assay*

The first and most ingenious qualitative test for isolating acrasin was introduced by Shaffer (1956a) who realised that although amoebae need to be covered by a film of water, an underwater technique has the disadvantage that the process of adding an active extract disturbs any existing diffusion gradient. He devised a test in which the amoebae were sandwiched in a film of water between agar and a glass surface (Fig. 2). Gradients of chemicals could be maintained owing to the absence of disturbing water currents. He cut out a block of agar carrying amoebae that had started to form streams, and turned it upside down on a slide. Gentle pushing of the agar block freed the amoebae wedged between the agar and the glass from the early streams. The experiment was carried out in a moist chamber to prevent evaporation of water and drying out of the agar. Wash-water from pseudoplasmodia was filtered through a cellophane membrane to remove amoebae and repeatedly

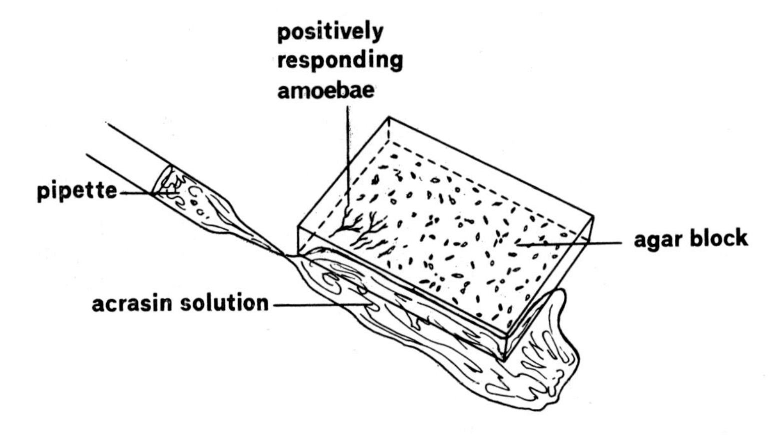

FIG. 2. An acrasin solution has been repeatedly placed at one side of the agar block. Some of the amoebae which are sandwiched between the agar block and a glass slide have responded positively.

carried over to the agar block in a pipette or transferred across the slide with a glass rod, so that the acrasin-containing water was applied to one side of the agar block. The reaction was considered positive when a large number of amoebae became oriented perpendicularly to the edge (Fig. 2). The active molecules were clearly small, since they had passed through a cellophane membrane. Unfiltered wash-water lost its chemotactic effect when stored at 22°C for twenty minutes but retained its activity if filtered through a membrane. This was due to the presence of an acrasin-inactivating enzyme, later identified as a phosphodiesterase (Chang, 1968), in the wash-water of the pseudoplasmodia.

C. The flow assay

Francis (1965) made Bonner's underwater flow technique more sophisticated by regulating the depth of the water and the rate of flow (Fig. 3). Water flowing over an aggregate resulted in the formation of a tail downstream from the centre, and large centres produced longer tails than small ones. The density of the tail reflected not only the strength of the centre but also the sensitivity of the responding amoebae. Exposure of amoebae of *Polysphondylium pallidum* of different physiological ages to centres led to a sudden rise in sensitivity when the amoebae were close to the aggregative phase. The amoebae also responded in the flow assay when wash-water of amoebae was concentrated and continuously injected into the flow chamber. The flow test, although more quantitative

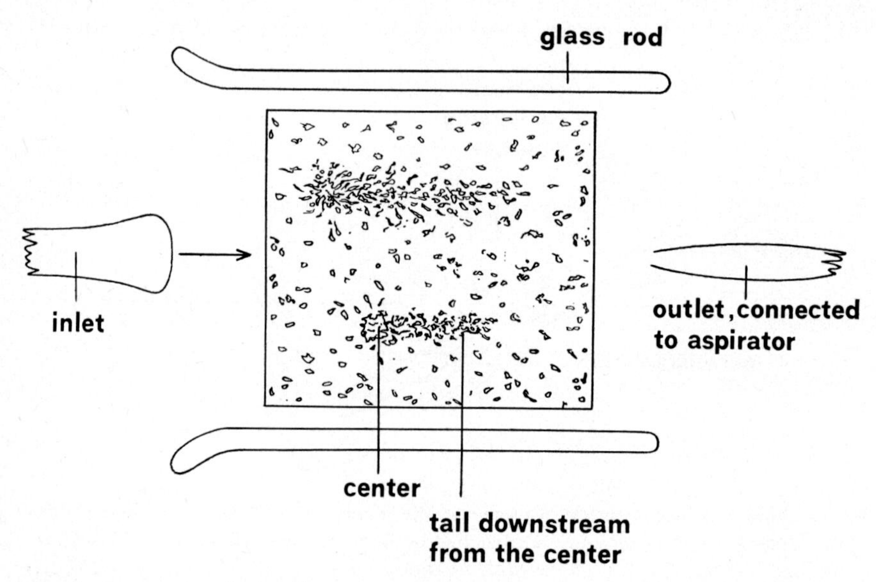

FIG. 3. An acrasin solution flows over amoebae on a coverslip. Amoebae downstream of the centre react positively, resulting in the formation of "tails". Redrawn from Francis (1965).

than the sandwich test, does not permit distinction of a response to the centre from a response to down-stream cells that may become secondary centres of attraction.

D. *The cellophane square assay*

Bonner *et al.* (1966) allowed amoebae in suspension to settle on small squares of cellophane, which were then placed in small petri dishes containing plain agar or agar mixed with an active extract. The amoebae crossed the edge of the cellophane membrane (Fig. 4) and the distance between one edge and the four amoebae which moved furthest away from it was measured after 1, 2 and 4 hours. Outward movement on agar mixed with a solution of concentrated wash-water of aggregating amoebae is greater than on plain agar.

The outward movement was independent of the physiological age and density of the amoebae. When very active solutions derived from the bacterium *Escherichia coli* were added to the agar the amoebae moved out at a speed of 1·0 mm/h. Dilution of the active material produced a gradual decrease in the rate of movement. On plain agar the speed was 0·3 mm/h. Since the active factor increased the rate of movement Bonner *et al.* (1966) called it the "rate substance". This active agent was heat stable, non-dialysable (it was later shown to be bound to a polysaccharide) and became inactivated when stored at 22°C, which suggests that an inactivating enzyme is present. Assays of *D. discoideum*, *D.*

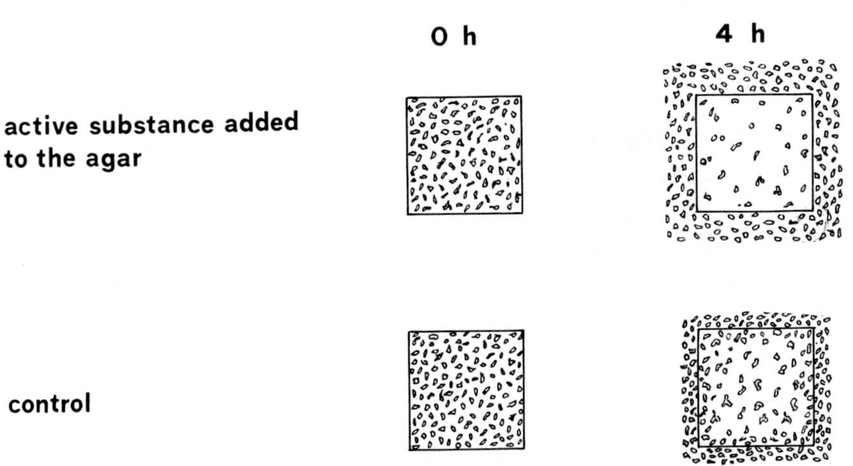

Fig. 4. An active substance has been added to the agar and a cellophane square on which amoebae were allowed to settle has been placed on the agar. The amoebae migrate from the square more rapidly than in the control. Redrawn from Bonner *et al.* (1966).

mucoroides and *P. violaceum* gave no indication that the amoebae produced or responded to species-specific substances. Yeast extract as well as bacteria was a good source of the "rate substance".

Later, using time-lapse films, Bonner *et al.* (1970) measured closely similar rates of movement on plain agar and on agar containing $1{\cdot}5 \times 10^{-5}$ M cyclic AMP or wash-water from intact *E. coli*, but only amoebae on the threshold of aggregation showed significantly increased activity when cyclic AMP was added to the agar. They concluded that the increased net rate of outward movement of the cells under these conditions was caused less by an increased rate of movement of individual amoebae than by an oriented movement outward. Phosphodiesterase secreted by the amoebae was thought to hydrolyse the cyclic AMP surrounding the amoebae, creating a gradient which would orient the amoebae away from the cellophane square.

A second bacterial factor derived from the *E. coli* wash-water induced not only an increase in net outward movement but also an increased rate of movement of individual amoebae. A bacterial extract from which cyclic AMP had been removed by dialysis was highly active with vegetative amoebae but had no effect on aggregating amoebae of *D. discoideum*. The most likely candidate for this second chemotactic factor appeared to

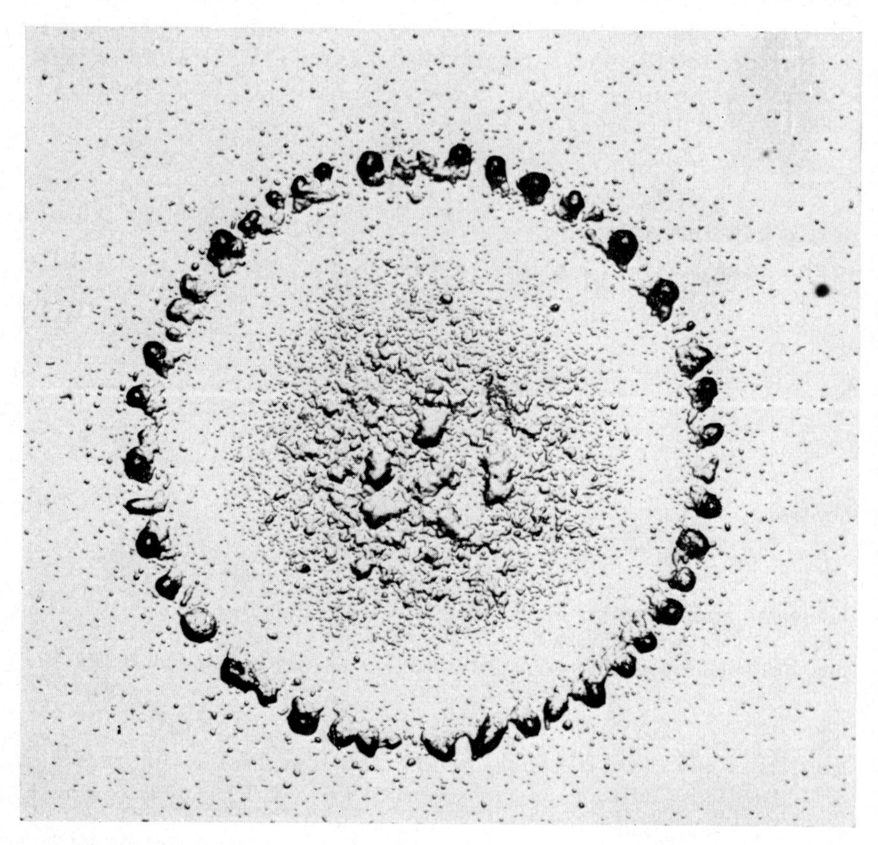

FIG. 5. Amoebae have been placed on agar containing 0·1 mg of cyclic AMP per ml. The ring has been caused by the centrifugal movement of the amoebae. From Konijn *et al.* (1968).

be folic acid or one of its derivatives (Pan *et al.*, 1972). The non-diffusibility of this second bacterial factor was thought to be due to the binding of folic acid to a polysaccharide.

It seems clear, therefore, that in this assay crude bacterial extracts can be effective either through containing cyclic AMP, which has acrasin activity, or through the presence of a second factor, folic acid.

E. The Boyden assay

The Boyden assay (see also Chapter 6) was designed to test chemotaxis of leucocytes. A chamber is separated into two compartments by a millipore filter, and the suspension of test cells is placed in one compartment and a solution containing the test substance in the other. The gradient of the test substance across the millipore filter attracts cells into the filter. The number of cells penetrating through the filter is a measure of the activity of the test substance. Bonner, Hirshfield and Hall (1971) found the optimal pore size for amoebae of *D. discoideum* to be 1·2 μm. The test was run for 1·5 h at 21°C in constant light. The amoeba concentration used was about 2,500 per ml. The amoebae are counted easily after fixation and staining.

The results obtained with the Boyden assay are consistent with the information previously obtained from the cellophane-square test. In both tests 10^{-7} M cyclic AMP and bacterial extract are active and the latter is the more active of the two. The main advantage of the Boyden assay is the elimination of the variation caused in the cellophane square test by changes in humidity and degree of dryness of the agar surface.

In both the cellophane square and Boyden assays the amoebae form a dense ring of cells similar to that illustrated in Fig. 5. This uneven distribution of the cells is more pronounced close to the aggregate stage and might be attributable to an inactivating enzyme (Bonner, Hirshfield and Hall, 1971).

F. The small-population assay

The small-population test requires a hydrophobic agar surface of a specific rigidity on which small populations of pre-grown amoebae are deposited. Agar becomes hydrophobic after repeated washing with de-ionized water (Ennis and Sussman, 1958); the washed agar is then boiled and allowed to gel. The surface is then water-repellent. Such an agar is very hypotonic, and the lack of salts harms amoebae. When the salt content is restored before gelation, the agar surface remains hydrophobic and is not detrimental to the amoebae (Konijn and Raper, 1961). The amoebae of a small population stay within the drop and with *D. discoideum* form only one aggregate, which permits investigation of the kinetics of cell aggregation without interference from neighbouring aggregates. The amoebae move freely within the boundaries of the drop and there are no convection currents. In the case of these small populations it is possible to show that the presence of special cells is not obligatory for aggregation in *D. discoideum*. The hydrophobia of the surrounding agar surface prevents the amoebae from crossing the boundaries of the droplet (Fig. 6) except at agar concentrations of 0·3% or less, when the

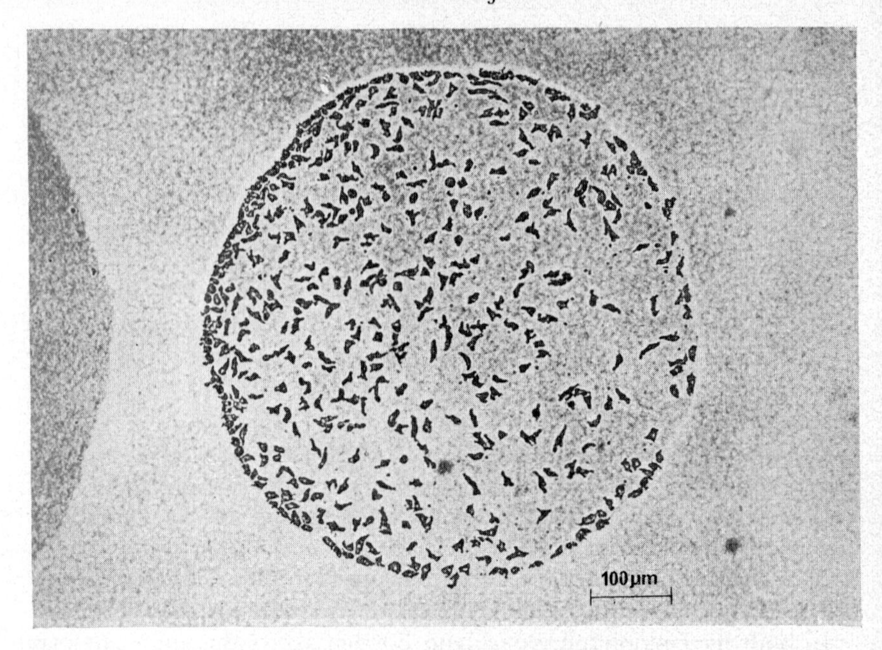

FIG. 6. Drops (0·1 μl) containing 3×10^{-13}g cyclic AMP were deposited three times at 5 min intervals to the left of the responding drop. Amoebae within this drop have been attracted but have not crossed the boundary. From Konijn (1970).

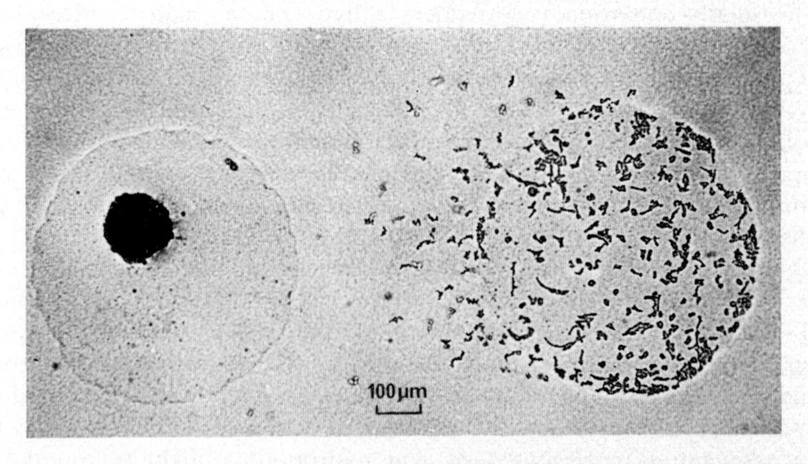

FIG. 7. Chemotaxis in *Dictyostelium discoideum*. The aggregation of amoebae in the drop on the left attracts amoebae which cross the boundary of the drop on the right. From Konijn (1970).

surface becomes too weak to hold the amoebae and cells move into the agar and outside the margins of the drop. At a concentration of 0·5%, all amoebae of *D. discoideum* stay inside the drop except when attracted by aggregating amoebae in a neighbouring drop (Fig. 7). Other species require a slightly higher agar concentration to keep the cells within the boundaries of the drop. Hydrophobia of the agar is a pre-requisite for the assay, because on a normal, hydrophilic agar the amoebae do not remain confined within the drop. The amoebae used in the assay are grown prior to use on nutrient agar in association with bacteria.

The small population assay permits the study of chemotaxis in amoeboid organisms in several ways:

1. *Amoebae or bacteria as an attracting source*

(a) The effect of environmental conditions such as light, temperature, and cell density on chemotaxis have been studied with amoebae as the chemotactic source. The attracting amoebae are deposited some time before the responding amoebae in the neighbouring drop so that the latter cells are sensitive at a time when the first drop contains aggregating cells (Fig. 7). Each drop has a volume of 0·1 μl and its diameter after deposition is about 0·6 mm. One petri dish contains 100 to 150 drops, each with 500 to 1,000 amoebae, and the responding drops are placed at various distances from them. The strength of the chemotactic activity is estimated by increasing the distance between the attracting and the responding drop and determining the distance at which 50% of the responding populations react positively. The distance of attraction is measured between the centre of the aggregate at the time all the streams have entered and the closest side of the responding drop. This is more satisfactory than measuring the distance between the edges of attracting and responding drops; the latter measurement does not yield a close correlation between distance and the percentage of positively responding drops. The method has been used to demonstrate attraction quantitatively with species that respond to cyclic AMP, species insensitive to this nucleotide as attractant (Konijn, 1972a), amoebae of myxomycetes (Konijn and Koevenig, 1971), other soil amoebae, and amoebocytes of sponges (Rasmont, unpublished).

(b) The rate of cyclic AMP secretion by bacteria has also been determined this way. However, the best results are obtained by measuring the closest distance between bacterial populations and responding drops (Konijn, 1969). The response of the amoebae depends on the developmental stage of the amoebae, and the state of the bacteria. With this procedure the time of observation is critical. An alternative is to leave the cultures so that the amoebae move through the agar until they reach the bacteria and consume them. After exhaustion of the bacterial supply the presence of amoebae in the bacterial drop is unambiguous. When large distances between drops are used or bacterial species secreting reduced levels of cyclic AMP, the bacterial populations are not eaten. To obtain results that can usefully be compared it is necessary to use bacteria of similar age and density. The latter can be measured in a vitatron spectrophotometer. *E. coli* and *Aerobacter arrogenes* proved to be the most active species and presumably secreted more cyclic AMP than other bacteria. All species of the genus *Bacillus* tested showed a lower level of cyclic AMP production (Konijn, unpublished).

2. *Extracts or cyclic nucleotides as an attracting source*

Instead of aggregating amoebae or bacteria, use has been made of amoebal and bacterial extracts as the attracting source. Studies with purified fractions of wash-water of bacteria led to the identification of cyclic AMP as the chemotactic mediator for cell aggregation in the larger *Dictyostelium* species. Since vegetative amoebae are less sensitive than aggregating amoebae to this cyclic nucleotide (Bonner *et al.*, 1969; Konijn, 1969), only amoebae nearing the aggregation phase should be used as responding cells.

(a) The chemotactic activity of extracts and cyclic nucleotides is assayed not by varying the distance between attracting and responding drops but by using serial dilutions of the test substances to determine the dilutions at which attraction is exerted on responsive populations. The test substance is applied 300 μm from the drop containing the sensitive amoebae. Chemotactic activity is estimated by determining the dilution of the test substance that gives a positive response with 50% of the responding drops and the cyclic AMP dilution that also induces a 50% positive response. Amoebae of any species reacting to cyclic AMP could be used as responding cells, since they all react to cyclic AMP concentrations below 10^{-7} M. Normally, amoebae of *D. discoideum* are used because they are slightly more sensitive to cyclic AMP than the other species. Pre-grown cells are deposited and kept at 22°C for 2 to 3 hours, after which they are stored overnight at 5·5°C. On the following day the dishes are kept at 16°C for 30 minutes to acclimatise before the assay is performed at room temperature. The chemotactic response can be enhanced by repeated deposition of the active substance. By means of three applications of the test solution at 5-minute intervals close to the sensitive amoebae and observation of the response 5 minutes after the last deposition, the sensitivity of the response was increased tenfold. This bio-assay can be used for testing materials, such as bacterial extracts, milk, or urine, that are chemotactically active for species that do not respond to cyclic AMP, with a view to the isolation and identification of the active fraction.

(b) The assay is more sensitive if the criterion used for a positive response is the presence in the responding drop of at least twice as many cells pressed against the margin closest to the attracting drop than occur on the far side (Fig. 6). Cyclic AMP causes responding amoebae to press against the side at concentrations as low as 10^{-9} M, enabling the cyclic AMP content of various biological materials to be estimated (Konijn, 1970; Herrmann-Erlee and Konijn, 1970; van den Veerdonk and Konijn, 1970; van Wijk and Konijn, 1971). Twofold differences in cyclic AMP concentration can be shown in this way (Konijn, 1970). The test is applicable to crude extracts and only small volumes are needed, but it is unsuitable for the demonstration of small differences in cyclic AMP level. Cyclic AMP is the only natural compound shown to be active at such low concentrations for pre-aggregative amoebae of *D. discoideum*. After column and paper chromatography, the chemotactively active product from a range of attractant materials gave in various solvents the same Rf values as commercial cyclic AMP (Konijn *et al.*, 1969a). The only other naturally occurring cyclic nucleotide is cyclic GMP, the threshold of activity for which is 1,000 times higher than for cyclic AMP (Konijn, 1972a). The possibility that contamination of crude extracts by folic acid or its derivatives, which attract amoebae of several species (Pan *et al.*, 1972) is affecting results can be excluded as these compounds do not induce a positive response among starving amoebae of *D. discoideum*.

The threshold of 10^{-9} M refers to the assay, and not to the amoebae, which react to much lower concentrations of cyclic AMP. Most of the cyclic AMP molecules in the attracting drop diffuse into the agar in directions other than those in which sensitive amoebae are located. If it is assumed that cyclic AMP diffuses from the drop into the agar as a hemisphere of increasing radius, then at the threshold concentration of 10^{-9} M, in standard assay conditions and with the distance between the cyclic AMP drop and the amoebal population 300 μm, it can be estimated that less than 0·01% or 10^{-17} g of the cyclic AMP molecules will reach an individual amoeba. Narrowing of the distance between attracting and responding drops and application of smaller drops would give an even lower threshold for amoebal sensitivity. The kinetics of cyclic AMP diffusion in agar are complicated. The molecules diffuse not only down into the agar but also upward to places where the cyclic AMP concentration has been reduced because of hydrolysis by phosphodiesterase.

At higher concentrations (10^{-8} to 10^{-6} M) of cyclic AMP the amoebae are attracted out of the responding drop at one side and 10^{-4} M induces them to cross the boundaries at all sides and prevents aggregation. The reason for the outward movement of the cells, as in the cellophane square test, is probably that phosphodiesterase produced by the amoebae inactivates the cyclic AMP in and near the responding drop resulting in a cyclic AMP gradient. This would induce cells to cross the boundaries at all sides and move away from the margin over distances of as much as 0·8 mm. At still higher concentrations of cyclic AMP (10^{-2} M) the amoebae form small clumps of cells that move but remain inside the drop (Konijn, 1972a).

3. *Amoebae or solutions in slide imprints as an attracting source*

The edge of a microscope slide is pressed against hydrophobic agar without breaking the surface of the gel, and then another imprint is made at an acute angle to the first. These imprints leave two troughs about 0·5 mm deep, one of which is filled with test amoebae in the sensitive stage and the other with a chemotactically active solution or aggregates of amoebae. The maximal distance between the troughs over which a positive response is obtained is a measure of the activity. A response is positive when amoebae cross the edge of the trough and move into the agar (Konijn, 1972b).

III. Chemotactic factors—cyclic AMP and folic acid

Olive (1902) realised that the mechanism underlying chemotaxis could not be understood before the chemotactic agents were identified. Shaffer's sandwich assay led to a renewed search for the attracting compounds. Sussman *et al.* (1956) collected wash-water of aggregating amoebae of *D. discoideum* and kept their active preparations stable at a pH of 3·5. Paper-chromatography of a fluorescent fraction from a cellulose column gave two fluorescent bands which were both inactive in Shaffer's test but attracted sensitive amoebae when combined. Later R. R. Sussman and co-workers (1958) isolated three fractions which only induced chemotaxis in *D. discoideum* when they were combined. In a different ratio these same fractions were active for *P. violaceum*. It is difficult, however, to interpret these data in the light of later results because the fluorescent fractions contained a wide variety of compounds.

Identification of the chemotactic signal seemed imminent when the urine of pregnant women was found to be a potent attracting agent (Wright and Anderson, 1958). Steroids were tested for chemotactic activity by Shaffer's test, and progesterone, testosterone and oestradiol induced a chemotactic response. Shortly afterward, Heftmann *et al.* (1960) isolated an active sterol, Δ^{22}-stigmasten-3β-ol, from *D. discoideum*, and its presence in amoebae was confirmed by Ellouz and Lenfant (1971). A synthesized sample of this sterol showed some activity, but its importance as an acrasin became doubtful when the crude preparation proved to be a much stronger attractant than the purified product. Since steroids and alkaloids are surface-active agents, Heftmann *et al.* suggested a mechanism of aggregation by which the surface tension of the amoebae is lowered on the side closest to the source of the attractant leading to the extrusion of pseudopodia on this side, followed by oriented movement. Hostak and Raper (1960) discovered that alkaloids induced cell aggregation in *Acytostelium leptosomum*. Some steroids also evoked aggregation in this species.

The response of amoebae may be partially dependent on the assay used to evaluate chemotaxis. Active steroids such as oestradiol, progesterone, and testosterone do not score more than a 50% positive response at any concentration in the small population test with *D. discoideum*. Concentrations of 10^{-3} M and higher cannot be tested, because at these concentrations the steroids are insoluble in water. A natural acrasin should be effective at very low concentrations. The attractant secreted by only one or a few amoebae diffuses in all directions and very few molecules will reach the responding cell. If low concentrations of a compound do not attract amoebae in the small population test, it is unlikely that it is a specific acrasin secreted by amoebae to attract neighbouring cells.

Shaffer (1956b) showed that the molecule involved in the attraction of amoebae by aggregates was small, heat stable, and inactivated by an enzyme. The purification and identification of the attractant was carried out with a more quantitative assay (Konijn *et al.*, 1967). With the intention of purifying acrasin the small population assay was first applied to dialysed attractants of amoebae. Aggregates were covered with dialysing membrane and drops of water were placed on the membrane above one or more aggregates. At various times, the drops into which acrasin had supposedly diffused were tested, also in concentrated form, to determine the chemotactic activity. Since the weak activity obtained this way did not seem promising for continued purification, bacteria were used as a stronger source of attractant. Dialysed aqueous bacterial extract was purified by paper chromatography (with the help of J. G. C. van de Meene of the University of Utrecht). When pooled and concentrated active fractions were purified by paper electrophoresis at pH 3·9 the active molecule proved to be negatively charged.

To increase the yield of attracting substances, the aqueous bacterial extract was used directly instead of the dialysed products. *E. coli* was grown in large trays, washed, and centrifuged. The concentrated supernatant was highly active, and after fractionation by gel-filtration, paper chromatography and paper electrophoresis the purified product still induced a strong chemotactic effect.

Meanwhile, Bonner and co-workers had started to purify bacterial fractions using the cellophane-square assay. In 1967 we pooled our efforts at Princeton University, where we continued with the small-population test because it gave a specific response to a substance with a molecular weight of 200 to 400. The highly active product purified in

Utrecht gave a UV absorption spectrum with a peak at 259 nm (Fig. 8), which strongly suggested that the attracting compound was a derivative of adenine. The information we had already collected about the attracting molecule, and the presence of the attracting compound in bacteria and in urine, both known sources of cyclic AMP, led Barkley to suggest that we test commercial cyclic AMP. This cyclic nucleotide proved to be active at very low concentrations in the small population test and was

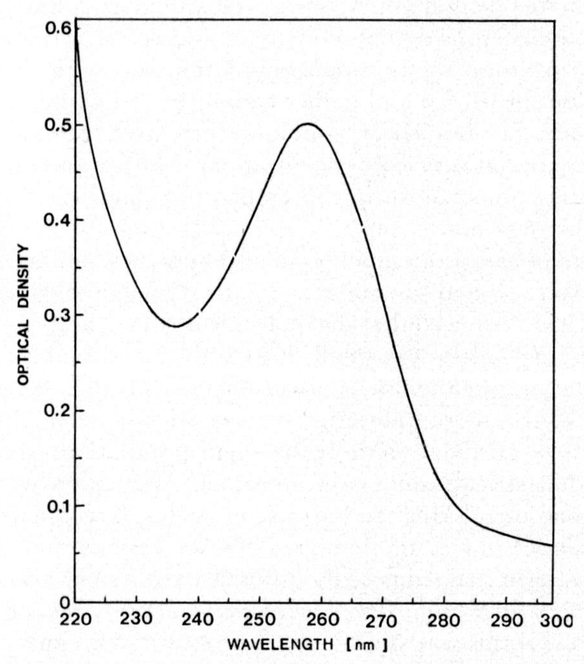

Fig. 8. Ultraviolet absorption spectrum of a pure chemotactically active fraction derived from *E. coli* extract, later shown to be cyclic AMP. From Konijn *et al.* (1967).

slightly active in the cellophane-square test, but other fractions containing large molecules were far more active in the latter assay. This activity was later found to be due to folic acid. Further purification on a DEAE Sephadex column and paper chromatography in three different solvents together with commercial cyclic AMP showed conclusively that the active compound from bacteria was cyclic AMP (Konijn *et al.*, 1969a).

It seemed possible that cyclic AMP also mediated aggregation of amoebae, since amoebae close to aggregation were most sensitive to it. The demonstration of the production of cyclic AMP by amoebae encountered some difficulties. It was relatively easy to show that amoebae of *P. pallidum* grown in shaken cultures on dead bacteria secreted a chemotactic compound that was identified as cyclic AMP (Konijn *et al.*, 1968). However, all fractions were tested for activity with amoebae of

D. discoideum. To our surprise, cyclic AMP secreted by *P. pallidum* did not attract amoebae of *P. pallidum*. Since amoebae of *D. discoideum* were sensitive to cyclic AMP, the attractant secreted by amoebae of the same species was purified. Initially, the yield of attractant was too small for identification. Chang (1968) explained this by demonstrating the presence of extracellular phosphodiesterase, an enzyme which could have hydrolyzed a large part of the cyclic AMP. Large quantities of amoebae of *D. discoideum* were grown on dead bacteria, and after centrifugation and resuspension were incubated in shaken cultures. After purification, the chemotactic substance secreted by these amoebae was identified as cyclic AMP (Konijn *et al.*, 1969b). A 3′, 5′-cyclic nucleotide phosphodiesterase inactivated the purified attractant completely. Using the same purification techniques and starting with dialysed products of the amoebae, Barkley (1969) also identified cyclic AMP as the attractant secreted by *D. discoideum*.

Other cyclic nucleotides were less active, and none of the analogues of cyclic AMP could induce chemotaxis at lower concentrations than cyclic AMP. The reaction of starving amoebae of *D. discoideum* to cyclic nucleotides is specific. Other nucleotides, sugars, amino acids and buffers varying in pH from 1·7 to 10·0 do not attract amoebae. It is probable that amoebae use this cyclic nucleotide during the vegetative stage to locate food and after starvation to mediate cell aggregation (Konijn *et al.*, 1968).

D. mucoroides, *D. purpureum* and *D. rosareum* also proved to be attracted by cyclic AMP (Konijn, 1972a). Binding of cyclic AMP to intact cells is limited to the species that respond to cyclic AMP. *P. pallidum* does not show binding activity (Mato and Konijn, 1975). Vegetative amoebae are also attracted by an additional factor (Bonner *et al.*, 1970). This factor is present in bacteria, yeast extract, milk, urine and casein, and absent in the lipid, polysaccharide, and protein fractions of caseins. Vitamin-free casaminoacids are also inactive. Of eight vitamins tested with the cellophane-square test, Pan *et al.* (1972) found that only folic or pteroylglutamic acid was active. Chromatograms of casein gave an active spot which was identified as folic acid. Folates are molecules whose structure is based on a pteroic acid skeleton conjugated with one or more L-glutamic acid molecules, and are present in plants, animals and micro-organisms.

The bacterial factor (page 107) which was a large non-dialysable molecule (Bonner *et al.*, 1970) proved to be a polysaccharide contaminant to which folic acid was bound. Vegetative and aggregative amoebae of seven species of *Dictyostelium* and *Polysphondylium* showed a chemotactic response to folic acid and dihydrofolic acid. The latter is a co-enzyme derived from folic acid by reduction with dehydrofolate reductase. Precursors of folic acid, for example biopterin and pterine, are also active in the cellophane-square assay. The test plates must be incubated in the dark to prevent photolytic degradation of folic acid.

The response of amoebae of *P. violaceum*, *P. pallidum* and *D. purpureum* was similar in the cellophane-square test over a concentration range of 10^{-2} M to 10^{-8} M; in other species optimum concentrations varied from 10^{-7}–10^{-8} M in *D. minutum* to 10^{-3}–10^{-4} M in *D. mucoroides*. Amoebae are attracted to folic acid before and after the onset of aggregation, which means that folic acid and its analogues might be acrasins for species that are not attracted by cyclic AMP.

Folic acid also attracts amoebae of seven *Dictyostelium* and *Polysphondylium* species tested in the small population assay (Konijn, unpublished). In most of these species the threshold activity of folic acid was 10^{-5} to 10^{-6} M, but in *D. minutum* it lay at still lower concentrations. The actual concentration active at the site of the amoebae of this species is of the order of 10^{-10} M. The threshold concentration for the response of aggregating amoebae in the small population test is two orders of magnitude higher than in vegetative amoebae. *D. aureum*, however, is more sensitive to folic acid in the aggregative stage. Amoebae of *D. discoideum* were not attracted by folic acid in the aggregation phase, and vegetative amoebae of this species showed a weak response.

The much higher concentrations of folic acid needed in the small-population assay for a chemotactic response by starving (aggregation phase) than by vegetative amoebae make it improbable that folic acid is an acrasin, as does the delayed response of amoebae to folic acid. The optimal response to cyclic AMP after three applications at 5-minute intervals occurred within 10 minutes of the last application, whereas the optimal response to folic acid was observed after 45 to 60 minutes.

A characteristic shared with cyclic AMP is the emergence of the cells at all sides of a responding drop at high concentrations of folic acid. If an inactivating enzyme is responsible for the centrifugal movement of the cells all these species must synthesize such enzymes.

Whether cyclic AMP or folic acid is the chemotactic agent functioning as a food-seeking mechanism may depend on the species. To settle this question, the concentration of cyclic AMP and folic acid secreted by the bacteria and the sensitivity of the various species to both compounds should be determined. The threshold concentration of cyclic AMP for chemotaxis is much higher during the vegetative stage than in the aggregative stage, which would make folic acid a more likely candidate as the bacterial attractant for feeding amoebae. However, when amoebae are surrounded by an abundance of bacteria chemotaxis is not essential for location of the food source. Amoebae would require guidance to their prey only when they are some distance from the bacteria and are therefore starving. It is in this kind of situation that the larger *Dictyostelium* species are most responsive to cyclic AMP and least sensitive to folic acid.

More research on chemotaxis during the vegetative stage is needed to decide the role of these two attractants, and possibly other chemotactic agents, in the detection of food.

IV. Cyclic AMP as a second and first messenger

In most organisms cyclic AMP acts within the cell in which it was generated, on cytoplasmic components or at the gene level (Robison *et al.*, 1971). A natural extracellular effect of this compound has only been shown in the Acrasiales. An extracellular effect in higher organisms cannot, however, be excluded, because physiological concentrations of cyclic AMP in the medium induce the proliferation of cultured lymphocytes (McManus and Whitfield, 1969).

In higher organisms hormones are the extracellular agents or first messengers, and the discovery of cyclic AMP has thrown light on the intracellular mechanism by which the action of several hormones is mediated. Sutherland and Rall (1958) identified cyclic AMP as a key mediator in the increase of glucose in the blood resulting from the action of adrenalin (epinephrine) on the liver. The lateness of the discovery of this cyclic nucleotide, essential for the regulation of many processes in animals, plants and bacteria, is explained by its low natural concentration (about 10^{-7} M) and the presence of the hydrolysing enzyme phosphodiesterase.

When it was found that several other hormones also increased the cyclic AMP levels in various tissues, Sutherland and co-workers developed the second-messenger hypothesis. They postulated that the hormones act as first messenger on the plasma membrane and that adenylate cyclase is stimulated as a result of hormone-receptor interaction. Cyclic AMP is then derived from ATP, the reaction being catalyzed by membrane-bound adenylate cyclase (Fig. 9). The cyclic nucleotide is subsequently hydrolysed to 5′-AMP or binds to another receptor, identified in higher organisms as a protein kinase. The ways in which cyclic AMP brings about its effects, which may be the production of a hormone or an enzyme, or an increase in permeability, are largely unknown.

Hormones can be defined as substances secreted by specific cells, which have a precise effect on certain other cells at extremely low concentrations. This definition of a hormone also holds for cyclic AMP during aggregation in *Dictyostelium*. Cyclic AMP is produced by certain amoebae and diffuses extracellularly to other cells where it has a specific effect. Hormones act at concentrations of about 10^{-8} to 10^{-12} M. The minimum concentration of cyclic AMP required to stimulate a responding cell is less than 10^{-12} M.

The mechanisms by which cyclic AMP acts on the cell is not known.

Marking of cells with ^{32}P- and ^{3}H-labelled cyclic AMP showed that cyclic AMP acts on a component of the cell surface and does not penetrate through the plasma membrane (Moens and Konijn, 1974). Autoradiography of whole and sectioned cells also provided evidence that cyclic AMP exerts its influence at the plasma membrane before one-sided pseudopod formation occurs and a positive chemotactic response can be observed. Neither of these techniques eliminates the possibility that traces of cyclic AMP enter the cell, but since amoebae are attracted by extremely low concentrations of cyclic AMP it is unlikely that enough cyclic AMP molecules could pass through the plasma membrane to

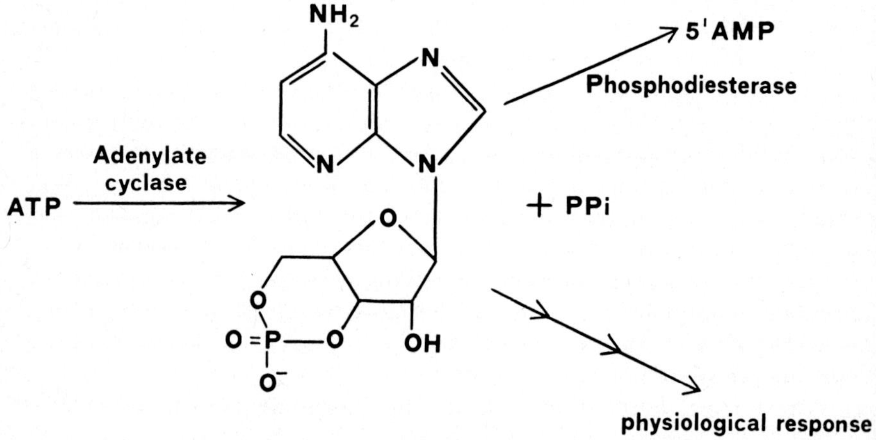

FIG. 9. The formation and breakdown of cyclic AMP.

activate directly a component within the cell that would be responsible for the positive chemotaxis. In animals, where it is assumed that the hormone binds to a receptor at the outer surface of a plasma membrane, the receptor may be the regulating sub-unit of adenylate cyclase or a part of it, perhaps a different part for different hormones. The regulating sub-unit is presumed to activate the catalytic sub-unit at the inner side of the plasma membrane which transforms ATP into cyclic AMP. Although hormones in mammals and cyclic AMP in *Dictyostelium* both bind to a receptor, it is unlikely that the receptor in the amoebae is a regulatory unit of adenylate cyclase. The stored energy in the 3′–5′ binding of the phosphorus group to the ribose, which is equivalent to that of ATP, may perhaps be utilised if cyclic AMP binds to receptors in the plasma membrane (Greengard *et al.*, 1969). An alternative is a re-cycling of cyclic AMP in which one cyclic AMP molecule activates several receptor molecules without being inactivated.

In the liver cell, a cyclic AMP-dependent protein kinase catalyzes the conversion of inactive phosphorylase kinase into its active form. Subsequently the active phosphorylase kinase converts inactive phosphorylase *b* into an active phosphorylase *a*, an enzyme which contributes to the breakdown of glycogen. It is known that cyclic AMP effects glycogenolysis, lipolysis, steroidogenesis, contraction, enzyme induction and permeability changes (Robison *et al.*, 1971). The reaction pathway followed after the activation of an amoeba by cyclic AMP is unknown. By analogy with other systems, one might expect cyclic AMP to react with a membrane-bound protein kinase responsible for the phosphorylation of proteins. Malkinson *et al.* (1973) measured the specific activity of cyclic AMP-binding proteins in *D. discoideum Ax*-2, and noticed a gradual increase in activity with time, the peak being reached after 18 hours, followed by a sharp decrease in activity. The amoebae aggregated after about 8–12 hours. The activity after 18 hours was seven to ten times higher than at zero time. The increase in sensitivity of the amoebae coincided with an increased activity of cyclic AMP-binding proteins. Weinstein and Koritz (1973) assayed protein kinase in intact cells of *D. discoideum Ax*-3. They, however, found no change in activity in the different stages of development and the activity was not stimulated by either cyclic AMP or dibutyryl cyclic AMP.

Whatever its action in cellular slime moulds and other organisms, a well-balanced regulation of cyclic AMP concentration is a pre-requisite for a harmonious interplay between the various processes under its control. Regulation takes place at various levels:

1. Regulation starts with the activation of adenylate cyclase, which is stimulated by hormones in higher organisms and by unknown factors in the cellular slime moulds. According to Shaffer (1958, 1962), the production of attractant is triggered by attractant secreted by neighbouring cells. Such stimulation is a pre-requisite for his "relay" hypothesis (see Section IX) and for *Dictyostelium* it would mean that cyclic AMP activates adenylate cyclase.
2. The continuing effectiveness of cyclic AMP requires its rapid breakdown, otherwise a high background level would nullify the effect of newly secreted cyclic AMP. Phosphodiesterase hydrolyses cyclic AMP to 5′-AMP, the activity of this inactivating enzyme changing with the concentration of cyclic AMP present. Methylxanthines can inhibit phosphodiesterase. The inactivating enzyme of *Dictyostelium* is less sensitive to methylxanthines such as theophylline and caffeine than that in other organisms (Chang, 1968). Natural inhibitors of phosphodiesterase are also known to occur but have not yet been identified.
3. Another way for the cell to rid itself of excess cyclic AMP is by excretion involving active or passive transport through the plasma membrane. Relatively large quantities of cyclic AMP are discharged by the kidneys and lactiferous glands. About half of the renal cyclic AMP has been filtered from the blood by the glomeruli, the rest coming

from the kidney tissues. Once outside the animal cell, cyclic AMP is probably merely a waste product. Leakage of cyclic AMP also occurs in bacteria. In *Dictyostelium*, too, cyclic AMP passes to the exterior, but instead of being a waste product it is essential for cell aggregation. The amoebae of the larger species of *Dictyostelium* probably discharge cyclic AMP periodically.

Some effects of cyclic AMP can be better understood by considering organisms lacking hormones, such as bacteria. Makman and Sutherland (1965) found low cyclic AMP levels when *E. coli* was grown in a glucose-rich medium, but when the glucose became exhausted the level rose rapidly. This was explained when the presence of glucose was found to reduce the activity of adenylate cyclase, and glucose also reduces the activity of many other enzymes. The advantage of this catabolite repression is that enzymes, the substrates for which are not available, are not synthesized. Even when a substrate is present, the breakdown of catabolites other than glucose can be repressed by glucose. In a medium containing lactose and glucose, α-galactosidase, an enzyme that splits lactose into galactose and glucose, will only be synthesized after glucose is exhausted and the synthesis of cyclic AMP is increased. Pastan and Perlman (1970) assumed that cyclic AMP was necessary for the transcription and translation of genetic information. Exhaustion of glucose results in activation of adenylate cyclase, and the cyclic AMP formed is thought to form a complex with a protein. The cyclic AMP-protein complex would then bind to, and activate, specific DNA, resulting in the production of specific proteins via mRNA. It would seem however, that a relatively small proportion, about 1% of *E. coli's* 10,000 genes, are regulated by cyclic AMP. Cyclic AMP also influences the formation of flagella in bacteria and hence bacterial movement (Yokota and Gots, 1970).

In eukaryotic organisms steroids as well as cyclic AMP may act in a way resembling that shown for cyclic AMP in bacteria; steroids too have been claimed to be chemotactically active in micro-organisms and also bind to specific proteins with the steroid-protein complex acting on specific DNA in mammals. In yeast (van Wijk and Konijn, 1971) catabolite repression is relieved after glucose depletion; the cyclic AMP level rises and enzymes become derepressed. Depletion of extracellular glucose does not trigger increased cyclic AMP production in the Acrasiales (Malkinson and Ashworth, 1972), but this glucose may not penetrate the plasma membrane and intracellular glucose might be exhausted while there is still an abundance outside the cell. The addition of cyclic AMP to cultured cells from malignant tumors slows growth and improves adhesion to the surface of the substrate. It gives them the appearance of normal cells; growth occurs in monolayers, and contact inhibition is restored. It should be emphasized that cyclic AMP, however, does not restore all of the characteristics of a normal plasma membrane.

In addition to its role in locating food and in cell aggregation, cyclic AMP plays a role in later developmental stages of *Dictyostelium*. After aggregation occurs, cyclic AMP may be one of the regulators of cellular differentiation since the application of 10^{-3} M induced the amoebae of *D. discoideum* to differentiate into stalk-like cells (Bonner, 1970). Bonner (1949) had already found that acrasin production was greater in the tip of the pseudoplasmodium than in the middle or posterior part. The anterior cells are the potential stalk cells of the fruiting body (see Fig. 1). This was confirmed by Garrod and Malkinson (1973), who found a higher cyclic AMP concentration in the tip. It seems unlikely, however, that cyclic AMP is the only regulator of the ratio between stalk cells and spores. If washed amoebae are deposited on filter paper and exposed to 3×10^{-3} M cyclic AMP 16 to 18 hours later, the normal polarity of the cells is disturbed and the cyclic AMP interferes with the synthesis and disappearance of some enzymes (Nestle and Sussman, 1972); the fruiting structures are abnormal at this high cyclic AMP concentration. Cyclic AMP has also been shown to influence differentiation in other organisms, for example, fruiting in *Coprinus macrorhizus* (Uno and Ishikawa, 1973).

By analogy with the effects of cyclic AMP in bacteria and higher organisms we may assume that it is also active inside amoebae. The relatively large amounts of this compound produced by cellular slime moulds that do not respond to cyclic AMP as a chemotactic agent indicates some internal function. With immunofluorescent techniques cyclic AMP has been shown to be uniformly distributed in the nucleus and cytoplasm of several members of the Acrasiales (Pan *et al.*, 1974). Other effects of cyclic AMP on amoebae include increased cell adhesion (Konijn *et al.*, 1968), centre inhibition (Konijn, 1969) and a reduction of territory size when added to the substrate in high concentrations (Konijn *et al.*, 1968).

V. Adenylate cyclase

Adenylate cyclase is membrane-bound and has the characteristics of a lipoprotein (Sutherland *et al.*, 1962), although it can be solubilised from *E. coli* (Tao and Lippman, 1969). The specific activity of this enzyme is constant throughout the life cycle of *D. discoideum* (Rossomando and Sussman, 1972). The yield of cyclic AMP was 3·4 nmol per min per mg of protein at 37°C. Malkinson and Ashworth (1972), who measured a peak in cyclic AMP concentration shortly before the highest level of phosphodiesterase activity was reached, assumed that the intracellular cyclic AMP concentration is regulated not by phosphodiesterase but by changes in the adenylate cyclase activity. Rossomando and Sussman (1973) solubilised adenylate cyclase and thought that the peculiar reaction kinetics of the enzyme (see Fig. 10) might explain the pulsations occurring during cell aggregation (see section IX). They found that the activation of adenylate cyclase depends on 5′ AMP. The ATP-pyrophosphohydrolase which converts ATP into 5′ AMP is activated by cyclic AMP, the derivation of which from ATP is catalyzed by adenylate cyclase. The adenylate cyclase of *D. discoideum* and *P. violaceum* has been

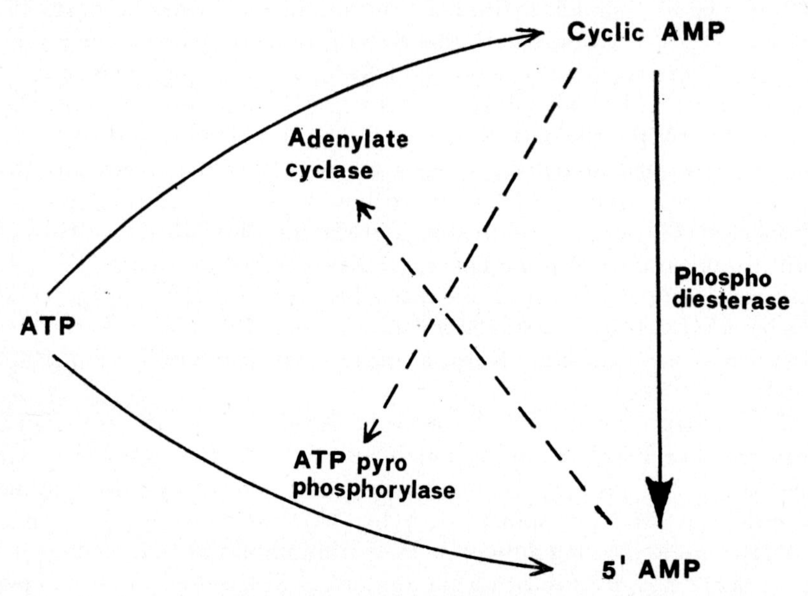

FIG. 10. The formation and breakdown of cyclic AMP, showing the activation (– – –) of the enzymes acting on ATP by their reaction products.

purified 50- to 100- fold and in both species the enzyme responded to 5′ AMP in the same way. Malkinson and Ashworth (1973) measured the cyclic AMP concentrations inside and outside the amoebae and found that the amount of adenylate cyclase was much higher than would be necessary for the rate of increase of intracellular cyclic AMP. They explained this discrepancy by suggesting a destruction of the structural integrity of the enzyme-membrane complex, which might change the kinetic properties of the enzyme.

VI. Phosphodiesterase

Before the identification of cyclic AMP as an attractant, Shaffer (1956a) showed that acrasin is inactivated by an enzyme. Such an inactivating enzyme would lower the concentration of attracting molecules and thus prevent an excessively high background level of acrasin and would also steepen the gradient of acrasin making it easier for the amoebae to respond. The production of extracellular phosphodiesterase, which catalyzes the hydrolysis of the 3′-bond of cyclic AMP, was demonstrated by Chang (1968). The phosphodiesterase activity in ruptured amoebae was only 4% of the level in the medium. The optimal pH was 7·5, and Mg^{2+}- or Mn^{2+}-ions were required for maximal activity. Chang measured a high Km of 2mM. Caffein, an inhibitor of phosphodiesterase in mammal tissues, did not inhibit the phosphodiesterase of *D. discoideum*, even at a concentration of 10 mM. The same concentration of theophylline gave only 15% inhibition, whereas it inhibits 90% of the phosphodiesterase activity of bovine heart (Butcher and Sutherland, 1962). Chang suggested a molecular weight of roughly 300,000. Chassy *et al.* (1969) also isolated a phosphodiesterase with a similar Km from *D. discoideum*. The enzyme hydrolysed cyclic CMP, cyclic TMP, and cyclic UMP, with similar Km values. Cyclic IMP and cyclic GMP were hydrolysed more slowly, and dibutyryl cyclic AMP was not hydrolysed at all. *P. pallidum* which does not respond to cyclic AMP by taxis does not have an extracellular phosphodiesterase (Konijn *et al.*, 1969b).

In liquid cultures phosphodiesterase activity increased during growth of amoebae of *D. discoideum* and declined within 2 hours of the consumption of the bacteria. In the aggregative stage only 2% of the original activity was left (Riedel and Gerisch, 1971). The inhibitor is secreted several hours before aggregation starts. The phosphodiesterase activity in a culture with vegetative amoebae was inhibited after the supernatant of a culture of starving amoebae was added, and the inhibition was proportional to the amount of supernatant. Further research on the inhibitor showed that it was not dialysable, was heat-stable, sensitive to trypsin, and had a molecular weight of 40,000 (Gerisch *et al.*, 1972). Several species which secreted phosphodiesterase also produced the inhibiting factor. With the exception of *D. purpureum* which was only partially affected by inhibitors of *D. discoideum* and *D. mucoroides* (Gerisch *et al.*, 1972), inhibitors produced by one species were fully effective for another species. In *D. minutum*, *P. pallidum* and *P. violaceum* which do not respond to cyclic AMP, the inhibitor is lacking. Except for *P. violaceum* they have little or no extracellular phosphodiesterase activity (Table I).

Natural inhibitors of phosphodiesterase occur elsewhere in nature. Shimoyama *et al.* (1972) isolated an inhibitor of small molecular weight from potatoes. The inhibitor isolated from *Dictyostelium* does not affect phosphodiesterase of bovine heart thus showing some specificity (Gerisch *et al.*, 1972).

Mutants of *D. discoideum* deficient in the production of the inhibiting factor produced greater amounts of active phosphodiesterase than the wild type. Riedel and Gerisch (1971) concluded that the inhibition of phosphodiesterase may be necessary for initiating the aggregative phase by increasing the external cyclic AMP concentration.

TABLE I. Cyclic AMP and phosphodiesterase activities in species of *Dictyostelium* and *Polysphondylium* (after Bonner *et al.*, 1972).

Species	Extracellular cyclic AMP	Phosphodiesterase Activity		Attraction to cyclic AMP
		Intracellular	Extracellular	
D. minutum	+	++	0	0
P. pallidum	+	++	0	0
P. violaceum	+	++	++	0
D. discoideum	++	++	++	++++
D. purpureum	+++	++	++	++++
D. mucoroides	+++	++	++	++++
D. rosareum	+++	++	+	++++

Riedel *et al.* (1973) distinguished four groups of morphogenetic mutants: those lacking or with a reduced phosphodiesterase activity, those producing an excess, those in which production of the inhibitor is absent or delayed and those with defects unrelated to phosphodiesterase activity. Since phosphodiesterase regulates the cyclic AMP concentration, defects in the regulation of phosphodiesterase activity lead to abnormal aggregation (Table II). A defect in inhibitor production results

TABLE II. Defects in cyclic AMP-phosphodiesterase regulation and in morphogenesis of mutants (after Riedel *et al.*, 1973).

PD regulation	Mutant no.	Morphogenetic aberrations	
		Aggregation	Fruiting body formation
1. PD activity almost absent	*ga* 86 *ga* 88	Extremely large aggregation territories, with marked streams of cells	Delayed formation of typical fruiting bodies

2. Excessive PD production	*aggr* 75	Cells aggregate without streams. Aggregation centres often develop into rings, which then break into separate cell masses	Small fruiting bodies
3. Inhibitor completely or virtually absent	*Wag*-4 *Wag*-7 *Wag*-10	Cells aggregate without or, in case of *Wag*-10, with few indistinct streams	Few small, atypical fruiting bodies
	aggr 50–2	Virtually none	None
	aggr 52	Small cell groups, no streams of aggregating cells	None
Inhibitor activity reduced	*Wag*-1	Cells aggregate without streams	Few small fruiting bodies
	fty 17	Cells show almost typical aggregation	At high cell densities, groups of irregularly intermingled spores and stalk cells in place of fruiting bodies (Sonneborn *et al.*, 1963)
Inhibitor production delayed	*Wag*-9	Cells aggregate without streams	Rudimentary fruiting bodies
	aggr 50–1	Rounded cell groups; no streams of aggregating cells	None
4. No apparent defect in PD regulation	*Wag*-2 *Wag*-3 *Wag*-8	Cells retain an homogeneous layer	None
	Wag-6 *Wag*-11	No aggregation, only wave-like propagating cell accumulations	None
	Wag-5 *aggr* 53	Aggregation without or, in case of *Wag*-5, with few indistinct streams	Few rudimentary fruiting bodies
	fty 1	No streams of aggregating cells (Sussman, 1955)	Fruiting bodies extremely small

in small fruiting bodies, and an excess of phosphodiesterase activity gives a similar effect. When phosphodiesterase is almost absent, the aggregation territories are very large.

Goidl *et al.* (1972) inhibited phosphodiesterase with specific antibodies. Amoebae treated in this way were unable to aggregate. After removal of the antibodies by washing of the amoebae, aggregation proceeded normally. However, Riedel *et al.* (1973) reported normal aggregation in strains with strong phosphodiesterase inhibition, which suggests that inhibition of phosphodiesterase and cell aggregation are compatible.

The phosphodiesterase isolated by Gerisch *et al.* (1972) had a molecular weight of about 60,000, which is much smaller than the values reported earlier (Chang, 1968; Chassy *et al.*, 1969). Its Km was also lower, 4 μM, which makes it more suitable as a regulator of the cyclic AMP level. This phosphodiesterase hydrolyzed at least 95% of the extracellular cyclic AMP in the growth phase of liquid cultures with a cyclic AMP concentration of 25–50 μM. Pannbacker and Bravard (1970) also reported a phosphodiesterase with a low Km of 15 μM. The two forms might be explained by a slow conversion of the phosphodiesterase with a low Km to a less active form with a high Km (Chassy, 1972). These two distinct forms of phosphodiesterase may be physiologically significant. The phosphodiesterase with a low Km could inactivate the excess of cyclic AMP present in the environment, and a gradual change to a phosphodiesterase with a Km as high as 2 mM would allow higher concentrations of cyclic AMP to exist at a later phase (Chassy, 1972). Other biological systems are known to have more than one phosphodiesterase, each with a different Km (Butcher and Sutherland, 1962; Brooker *et al.*, 1968).

Malchow *et al.* (1972) described a membrane-bound phosphodiesterase which was not inactivated by an inhibitor in the starvation stage of the amoebae. Furthermore, this membrane-bound phosphodiesterase was not inactivated when the *D. discoideum* inhibitor was added. Unlike the extracellular phosphodiesterase, its activity was low during the vegetative phase and increased during the interphase period before aggregation. Its activity is high when the amoebae are most sensitive to the chemotactic effect of cyclic AMP and this phosphodiesterase was originally thought to be a component of the chemotactic receptor mechanism. Pannbacker and Bravard (1972), who isolated the same membrane-bound phosphodiesterase, suggested that it might contribute to the building up of a cyclic AMP gradient during aggregation. In this way it would supplement or replace the extracellular phosphodiesterase. After cyclic GMP was found to be a better substrate for membrane-bound phosphodiesterase than had been expected from its chemotactic activity, it became clear that the membrane-bound phosphodiesterase is not identical with the chemotactic receptor (Malkinson *et al.*, 1973;

Malchow and Gerisch, 1973). Membrane-bound and soluble phosphodiesterases have also been detected in other organisms, for example the myxomycete *Physarum polycephalum* from which Murray *et al.* (1971) isolated one particulate and one soluble phosphodiesterase.

Circular ring-shaped centrifugal movement of amoebae (Fig. 5) was observed when amoebae were placed in the centre of a petri dish containing cyclic AMP dissolved in non-nutrient agar (Bonner *et al.*, 1969). When theophylline was added, the outward-moving rings of amoebae did not appear or were delayed. Apparently the weak inhibition by theophylline (Chang, 1968) of phosphodiesterase (thought to create a cyclic AMP gradient) was sufficient to interfere with the outward movement of the cells. *P. pallidum*, which does not respond by taxis to cyclic AMP does not show these rings (Bonner *et al.*, 1969).

Amoebae spread on agar containing a high cyclic AMP concentration (10^{-4} M) aggregate and differentiate in a normal fashion with territories of the same size as those in control dishes (Bonner *et al.*, 1969). The effect of the high cyclic AMP level is hence presumably overcome by high phosphodiesterase activity induced by the high cyclic AMP concentration (Konijn, 1972a). In bacteria (Aboud and Burger, 1971) and mammals (D'Armiento *et al.*, 1972) the phosphodiesterase level is higher after the addition of cyclic AMP.

Different phosphodiesterases and their well-balanced regulation may be necessary to keep the concentration of cyclic AMP around the individual amoebae below saturation level. This would be necessary so that amoebae could detect the side with the higher cyclic AMP level and thus move towards the source of the chemotactic molecules.

VII. Mechanism of aggregation

It is generally assumed that the mechanism of aggregation in *D. discoideum* depends on a relay system (Shaffer, 1962; Gerisch, 1968; Cohen and Robertson, 1971a, b). The aggregate releases cyclic AMP in pulses, and amoebae responding to the periodically secreted attractant move to the aggregate. Activated cells secrete attractant that will activate more distant amoebae. These amoebae, before moving centripetally, secrete cyclic AMP which attracts amoebae still further away. A refractory period follows in which amoebae do not react to cyclic AMP. The production of cyclic AMP by amoebae in the refractory period decreases reaching a level which is too low to influence other amoebae. After a few minutes, cyclic AMP production rises and passes the threshold concenration again, resulting in the next pulsation. A concentrated zone of cyclic AMP propagates outward, and is visualized by time-lapse cinematography as pulsating movements of amoebae towards the centre. Rhythmic waves are not limited to cyclic AMP-sensitive species. The insensitive species *D. polycephalum* and occasionally *P. pallidum* (Raper, personal communication) also show pulses during aggregation. *D. discoideum*, *ap*-66, a mutant which is sensitive to cyclic AMP, aggregates without pulses (Gerisch *et al.*, 1966).

The interval between two pulses is generally about 5 minutes (Shaffer, 1962) in *D. discoideum*, *D. purpureum* and *D. mucoroides* but may be as short as 2 minutes or as long as 10 minutes. Some mutants of *D. discoideum* have abnormal wave periodicity (Durston, 1974). The interval in *D. rosareum* is 20 minutes (Konijn, 1972a). At an advanced aggregative stage the interval between two pulses tends to become shorter. Each inward pulsation of amoebae is the result of an outward-moving stimulus. The cells around the attracting centre react periodically and synchronously to cyclic AMP which affects the orientation and speed of the amoebae (Gerisch *et al.*, 1966). While orientation and speed normally reach their maximum at the same time, a phase shift in these two factors takes place in *D. discoideum*, mutant 39 (Gerisch *et al.*, 1966). Oscillatory movements controlled by cyclic AMP are present in cell suspensions of aggregating and of preaggregating cells (Gerisch and Hess, 1974). Analysis of the effect of cyclic AMP shows that the oscillations contain a fast and a slow component. Gerisch and Hess indicated that only the fast component has characteristics of the chemotactic response to cyclic AMP.

Nanjundiah and Konijn (unpublished) found the interval between two pulses in aggregating amoebae to be temperature dependent. Apparently, the endogenous biochemical processes underlying the rhythmic release of attractants slow down at lower temperatures, resulting in a longer period between two emissions of cyclic AMP. Oscillations in the concentration of cyclic AMP occur in sarcoplasmic reticulum (Gillibrand, 1970). An increase in the adenylate cyclase activity and a decrease of phosphodiesterase activity would lead to oscillation peaks. Spangler and Snell (1961) argued that interaction of two enzymes and their products could lead to continuous oscillations. For a more detailed discussion of oscillations in *Dictyostelium*, the reader is referred to Cohen and Robertson (1971a, b) and Gerisch and Hess (1974).

Further investigation is required to determine what change is induced by cyclic AMP in the plasma membrane or what other attributes of the amoebae are modified. Possible changes proposed by various authors include a gel-sol transition, changes in surface tension, and increased permeability at the side of the source of the attractant resulting in one-sided pseudopodium formation.

A relay system would allow aggregates to derive their amoebae from large territories. Large aggregation territories have been observed when cultures are left undisturbed (Shaffer, 1958) or high cell densities are transferred (Gerisch, 1965, 1968). An alternative to a relay mechanism of aggregation is a direct attraction by a centre of all free amoebae within its territory (Arndt, 1937; Bonner, 1949). Control of all individual amoebae directly by a centre was later considered unlikely, since a centre was assumed to attract neighbouring cells only over a distance of a few hundred μm (Shaffer, 1957, 1958). However, centres are now known to attract over distances of 1·6 mm (Konijn, 1965, 1968). The response was markedly positive when cells crossed the edge of the responding drop, which involved penetration of the agar. When pressing of amoebae in the responding drop against the side closest to the attracting drop (Fig. 6) was taken as criterion for a chemotactic response, positive chemotaxis was observed over distances of more than 2 mm.

Pre-grown amoebae placed on agar or a millipore filter formed aggregation territories with a radius of 0·65 to 0·8 mm (Bonner and Hoffman, 1963; Konijn and Raper, 1966; Sussman and Sussman, 1969). Such a radius would be sufficiently small to permit direct attraction by the developing aggregate of all amoebae within an aggregation territory, without requiring a relay mechanism.

When a relay of cyclic AMP is not involved in aggregation, a territory has to be delimited at an early stage when only a few cells occupy the centre, otherwise other cells in the territory would start their own aggregates. Chemotactically attracted amoebae moving to the centre are less well placed than a centre for forming a gradient of cyclic AMP that will reach threshold levels for their neighbours. Another possible reason why cells undergoing attraction do not themselves become centres is that movement utilizes their ATP supply with a consequent reduction in cyclic AMP formation.

It has been shown that an aggregate consisting of a small group of cells secretes cyclic AMP maximally and that a 10-fold increase in the number of amoebae in the aggregate does not give increased production (see section XIII). This implies that a small aggregate could attract all cells in its territory. If aggregations attract single amoebae directly, one would expect species with larger aggregation territories to release a chemotactic substance with an effect over larger distances. Evidence that there is indeed a correlation between territory size and effective distance for chemotaxis will be given in section XII.

When the distance between the edges of attracting and responding drops was measured and compared with the frequency of positive responses, no close correlation was obtained (see page 111). A close correlation was however found when the distance between the centre of the aggregate and the closest side of the responding drop is measured. This is consistent with the cyclic AMP secreted by the centre determining the chemotactic response in the neighbouring drop and not with a relayed zone of cyclic AMP released from the edge of the attractant drop.

A possible shortcoming of the relay system is its potentially unlimited propagation (Konijn, 1973a). If all aggregates are formed simultaneously, competition among the various centres may confine the propagation of the relay to the separate territories. Under some conditions of illumination, however, only a few small aggregates are formed early and the rest of the cells develop their own aggregates one or more hours later (Konijn and Raper, 1966). If the stimulus were relayed from cell to cell, one would expect under these conditions that large aggregation territories would be formed. Instead, the size of the aggregates proved to be even smaller than when all aggregations appeared at the same time, despite the presence of large areas with solitary cells around the small aggregations. The spreading of the chemotactic stimulus was not obstructed by a sudden change in cell density or a difference in physiological age of the amoebae. Direct-marking of the territory by an early aggregate would avoid the problem created by a relay, in which the acrasin concentration is not related to the distance to the centre.

The large territories observed by Shaffer (1958) originated from amoebae grown with bacteria in undisturbed cultures. Therefore, the waste products or cyclic AMP secreted by the bacteria may have influenced the size of the aggregations. Furthermore, the large aggregates obtained were characterized by large centripetal streams which secreted cyclic AMP as did the centre (Bonner, 1947, 1949; Shaffer, 1962).

The large aggregations observed by Gerisch (1965, 1968), which were characterized by a constant velocity of the inward-moving amoebae, originated from sheets of cells. Since the amoebae touched each other, adhesive forces may have influenced the territory size. These sheets of amoebae in the aggregation are comparable with streams of amoebae, and a transfer of pulses in streams with the same cyclic AMP concentration everywhere may be different from the spreading of the stimulus among free amoebae.

To avoid the effects of interference by other aggregates, Konijn analyzed the aggregative behaviour of about 70 amoebae in one small population (unpublished results). The hydrophobic agar surface kept the amoebae within the boundaries of the drop. There was no evidence that amoebae close to a newly formed aggregate affected the movement of amoebae at the periphery of the drop. Some amoebae lying close to the newly formed aggregate moved outward before being attracted by the centre. Arndt (1937) also observed amoebae that seemed to crawl uninfluenced by the attracting source. A different physiological stage in the unresponsive amoebae could account for this. The threshold of the physiologically less advanced amoebae would initially be too high to detect the chemotactic substance. Cells in the small drop were also observed to move to places where there was no cell that could have relayed the chemotactic stimulus.

When a constant speed of the outward wave of a chemotactic stimulus occurs a relay system seems probable. The concentric zones of inward-moving cells are a visible expression of the signal, which moves with a speed of 43 μm per minute (Gerisch, 1965). A similar speed of signal propagation was measured by Robertson *et al.* (1972), who used artificial periodic pulses of cyclic AMP in a field of sensitive amoebae. The cell density they used was much lower, about 600 amoebae per mm^2, than that of the packed cells used by Gerisch.

Attraction of amoebae of *D. discoideum* does not necessarily depend on 5 minute pulses of cyclic AMP. Amoebae in a small population on a hydrophobic agar surface respond chemotactically to one pulse of cyclic AMP and to drops of cyclic AMP applied at intervals as short as 2 minutes or as long as 15 minutes. Bacteria, which do not secrete cyclic AMP periodically, also attract amoebae. A continuous flow of cyclic AMP attracts amoebae, as shown by Bonner's (1947) under-water experiments (see page 104).

VIII. Chemotactic activity of cyclic nucleotides and their analogues

Several analogues of cyclic AMP and other cyclic nucleotides have been synthesized but of these only cyclic GMP has been shown to occur naturally. In contrast to the situation in other organisms or cells, in *Dictyostelium* no other cyclic nucleotide or analogue was more active than cyclic AMP. These other cyclic nucleotides are only effective at 10^2 to 10^5 times the threshold concentration for cyclic AMP, so the latter is in all probability the only natural acrasin for *D. discoideum*.

The amoebae of *D. discoideum* are very suitable for the study of relationship between structure and activity in analogues of cyclic nucleotides for the following reasons.

(a) During the pre-aggregative phase amoebae are extremely sensitive to cyclic AMP.
(b) At this phase they respond by taxis only to cyclic nucleotides and their analogues.
(c) The attractant apparently does not have to penetrate the plasma membrane to be active, whereas in other organisms it probably has to do so.
(d) The amoebae, unlike cells from tissues, do not need to be subjected to a disaggregation treatment prior to testing.

The chemotactic activity of 32 cyclic nucleotides and their analogues was studied with the small-population assay (Konijn and Jastorff, 1973; Konijn, 1972a, 1973b, 1974). Substitutions were introduced in the phosphate, base and ribose moiety of the cyclic AMP molecule. The phosphate group was essential to chemotaxis. Replacement of oxygen at the 5′-ribose position by a methylene or an amido group resulted in a chemotactic response similar to that of cyclic AMP. Replacement of oxygen by a methylene group at the 3′-ribose position resulted in almost complete loss of the chemotactic activity (Konijn *et al.*, 1969a). Addition of a protruding group to the substitute at the 5′-position reduced the chemotactic activity by a factor of 10^4 to 10^5. The type of group attached to the amido group at the 5′-position did not have a strong effect on chemotaxis. Without a negative charge in the phosphate group, the chemotactic activity was low but still detectable. Since diastereomeres (stereoisomers that are not mirror images of each other) of one analogue induced different chemotactic responses, the stereochemistry of the molecule apparently plays an important role.

Slight changes in the base moiety may affect the chemotactic response strongly. The importance of these changes indicates that structural features of the base react in some way with the receptor mechanism of the amoebae. The more lipophylic analogues are not more potent chemotactic agents. If the active molecule does not penetrate through the plasma membrane, a lipophylic compound that penetrates the cell more rapidly would not necessarily enhance chemotaxis.

Malchow *et al.* (1973) showed that the binding of analogues of cyclic AMP by amoebae is fairly well correlated with the concentration of particle-bound phosphodiesterase and with the chemotactic activity of the different analogues. These correlations do not establish, however, that the particle-bound phosphodiesterase is the receptor molecule of the analogue, especially as with cyclic GMP there was no correlation between binding to phosphodiesterase and chemotactic activity (see section VII). Extracellular and particle-bound phosphodiesterase gave similar rates of hydrolysis of cyclic nucleotides and their analogues.

IX. The role of calcium during aggregation

Rasmussen and Tenenhouse (1968) found that several cellular processes depend on calcium ions as well as cyclic AMP. They postulated that ATP is converted into cyclic AMP, which regulates the release of Ca^{2+}.

Mason *et al.* (1971) found that aggregation of amoebae of *D. discoideum* requires a Ca^{2+} concentration of at least 10^{-6} M. Other observations, however, indicate that in the absence of salts, including calcium, aggregation still occurs but it takes longer before the amoebae start to come together (Konijn and Raper, 1961; Malkinson and Ashworth, 1973). Subsequent stages in the morphogenesis of *D. discoideum* proceed more slowly after the addition of Ca^{2+} to the medium. This retardation was not due to high osmotic pressure, because similar concentration of other ions did not reduce the speed of morphogenesis and K^+ even increased it (Takeuchi and Tazawa, 1955). To study the effects of ions the concentration must be measured accurately, because the same cations can have a stimulatory or an inhibitory effect depending on concentration. The pH of the medium also influences the effect of cations.

A possible function of Ca^{2+} is to increase adhesiveness. Removal of divalent cations by EDTA changes the surface charge density and the adhesiveness of amoebae (Gerisch, 1961; Gingell and Garrod, 1969). Tetracaine, which alters the binding and transport of Ca^{2+}, prevents aggregation at 10^{-5} M but does not change cyclic AMP secretion (Mason *et al.*, 1971). At a concentration of 10^{-4} M tetracaine (Konijn, 1972c) chemotaxis of amoebae is significantly reduced. Chi and Francis (1971) found an increased outflow of Ca^{2+} from amoebae after the addition of 10^{-4} M cyclic AMP. In other organisms too cyclic AMP alters Ca^{2+} levels by changing membrane permeability (Rasmussen, 1970). The permeability of the cell membrane for Na^+ and K^+ ions was, however, not affected by cyclic AMP in *D. discoideum*. Chi and Francis speculated that Ca^{2+} exchange is involved in the amoebal contractile system required for cellular locomotion. Another function of Ca^{2+}, which is known for mammals, could be the regulation of phosphodiesterase activity (Kakiuchi, 1971). Gregg and Nesom (1973) studied the influence of cyclic AMP and Ca^{2+} on the plasma membrane. They concluded that cyclic AMP mobilized the intracellular Ca^{2+}, which altered the properties of the plasma membrane. The regulation of the Ca^{2+} level is so important in cell metabolism that Maeda (1970) proposed that other ions achieve their effects by controlling the intracellular concentration of Ca^{2+} ions.

X. Chemotaxis in cyclic AMP insensitive species

Cyclic AMP is the mediator of cell aggregation in *Dictyostelium* species with large fruiting structures, e.g. *D. discoideum*, *D. mucoroides*, *D. purpureum* and *D. rosareum*. All these species respond to 10^{-6} to 10^{-8} M cyclic AMP, and their aggregation is delayed or prevented at high concentrations of cyclic AMP.

The cyclic AMP-insensitive species release cyclic AMP but are not attracted by it. The smaller Dictyostelium species, e.g. *D. minutum*, *D. lacteum*, *D. aureum*, *D. polycephalum* and *D. vinaceo-fuscum*, are attracted by small, heat-stable molecules, similar in size to cyclic AMP, which are present in sources from which cyclic AMP has also been isolated. Bacteria, liver homogenates, urine and milk attract amoebae of some of the cyclic AMP insensitive species over larger distances than they do cells of *D. discoideum* (Konijn, 1972a). Another source of attractant is fluid in which amoebae have been incubated. Identification of the unknown attractants is hampered by loss of activity as purification of extracts proceeds. This could mean that chemotaxis depends on a synergistic effect of two or more agents or on the net effect of several independent attractants. When aggregation is not dependent on just a single attractant it becomes difficult to pinpoint the active components. Possibly, however, all species insensitive to cyclic AMP have only one chemotactic agent but we need more sophisticated techniques for identification. The effectiveness of the chemotactic system in cyclic AMP insensitive species becomes clear in the small-population test in which attraction occurs over distances of 2·5 mm, or 250 times the diameter of the amoebae. The attraction of amoebae to bacteria shows that all species are attracted by their food source. The binding of attractant to activated charcoal indicates that it could be related to cyclic nucleotides, although other chemicals are also adsorbed to activated charcoal.

After purification of *E. coli* extracts by column chromatography, charcoal treatment, and thin-layer chromatography with H_2O: NH_4OH:isobutyric acid (33:4:63 v/v) as solvent, the Rf value of the active component for *D. aureum* was 0·75, which is similar to that of cyclic AMP. In another solvent, *n*-butanol:acetone:acetic acid:5% NH_4OH:H_2O (35:25:15:15:10 v/v), the Rf value of the active fraction was 0·67 and that of cyclic AMP 0·56 (M. Tengbergen, unpublished). Cyclic AMP did not attract amoebae of *D. aureum*, whereas cyclic TMP and cyclic CMP induced a slight chemotactic response. The acrasin in cyclic AMP insensitive species might be related to folic acid, especially since amoebae secrete various substances similar to folic acid (Pan, Hall and Bonner; personal communication). B. Wurster in Bonner's laboratory obtained a highly purified attractant for *P. violaceum* that does not affect *D. discoideum* but it has not yet been identified. Size alone does not determine whether a species will be attracted by cyclic AMP, nor is the presence or absence of pulsations a valid criterion for sensitivity to it (Konijn, 1972a). Founder cells, which round off and initiate cell aggregation have been observed only in species that are not sensitive to cyclic AMP. However, one of these, *D. lacteum*, also lacks founder cells (Cohen and Robertson, 1972).

Cross attraction between amoebae of cyclic AMP sensitive species occurs. Mixed populations of these species enter the same aggregate (Raper and Thom, 1941). After a few hours the amoebae sort out, and the mixed aggregate falls apart into pseudoplasmodia of different species. Perhaps chemotaxis at first predominates over subtle differences in cell surfaces, but after some hours these differences lead to a separation into species. Another possible mechanism of sorting out is the occurrence of newly synthesized surface components after aggregation. These new compounds would cause specific differences between cell surfaces of different species, resulting in sorting out and splitting of the aggregate. As has been shown within a pseudoplasmodium, sorting out of cells may depend on cell density (Takeuchi, 1969; Bonner, Sieja and Hall, 1971).

Heavier cells take positions at the front and lighter cells move to the posterior part of the pseudoplasmodium. This difference in cell density does not exclude the possibility that cells in the front and back of a migrating pseudoplasmodium also differ in their adhesive properties. Besides cell density the culture medium and the developmental stage of the amoebae are considered determinative factors for sorting out (Leach *et al.*, 1973).

Amoebae in mixed populations of cyclic AMP insensitive species enter different aggregates. Cells may enter separate aggregates for various reasons:

(a) The attractant is species specific.
(b) A large difference in aggregation time results in aggregation in one species before amoebae of the other species become sensitive to an attractant.
(c) Aggregation in one species is inhibited by products secreted by the other species in the mixed population, e.g. certain species are sensitive to volatile inhibitors (see section XI).
(d) Amoebae of one species are attacked and engulfed by cells of the other species, e.g. when a small population of *D. discoideum* was mixed with *P. pallidum* or *P. violaceum* the amoebae of *Polysphondylium* were destroyed by amoebae of *D. discoideum* (Konijn, 1973a).
(e) Two species responding to the same attractant might nevertheless aggregate separately because of a difference in the periodicity of the aggregative movement.

In cyclic AMP sensitive species this last possibility does not seem to hold. *D. mucoroides* and *D. rosareum*, which differ in the duration of the interval between two pulses by as much as a factor of 4, enter the same aggregate and separate about 5 hours after the beginning of aggregation (Konijn, unpublished).

In order to study possible similarities in their acrasin, cyclic AMP insensitive species were mixed in ratios of 1:10; 1:1 and 10:1. Control drops contained only one species. Aggregates of different species in the same drop could be distinguished by differences in the size and morphology of the aggregates (Raper, 1960). The experiment was arranged so that aggregation would start simultaneously in both species. All of the cyclic AMP insensitive species we mixed (*D. aureum*, *D. lacteum*, *D. vinaceo-fuscum* and *D. polycephalum*) entered separate aggregates (Konijn, unpublished); hence it seems that species-specific mediators were involved in aggregation. No definite conclusions could be drawn about a specific acrasin for *D. polycephalum* because most of the amoebae encysted before aggregation. The formation of separate aggregates by mixed populations does not exclude a weak mutual attraction between different species. Amoebae of *P. pallidum* are attracted by aggregates of *D. minutum* and *D. mucoroides* responds to older centres of *P. violaceum* (Shaffer, 1957). The attraction of the larger *Dictyostelium* species to the cyclic AMP insensitive species is not surprising, since all of these species produce cyclic AMP.

Although *D. mucoroides* and *D. minutum* have many features in common, the difference in acrasin supports Raper's classification of *D. minutum* as a separate species (Raper, 1941). The insensitivity of *D. minutum* to cyclic AMP is not due solely to a difference in size. A small mutant of *D. mucoroides* of a size in the range of *D. minutum* was attracted by cyclic AMP (Konijn, 1972a). Other differences during aggregation of these two species are described by Gerisch (1964, 1966).

The occurrence of the unknown attractants in a large variety of sources, all of which contain cyclic AMP, suggests that, like cyclic AMP and folic acid, these unidentified compounds are functional in several other organisms. Urine and milk, which are rich sources of cyclic AMP, have to be diluted 100 to 1,000 times before their threshold activity is reached. Similar dilution was necessary to detect the lowest concentration at which cyclic AMP sensitive species would react to them.

Another similarity between the unknown attractants from various sources and cyclic AMP is the high degree of sensitivity when amoebae are nearing the time of aggregation. These attractants are evidently similar or identical to the chemotactic substances involved during aggregation in cyclic AMP insensitive species.

XI. Territory size and chemotaxis

The regulation of the size of an organism is a problem of great interest but of such complexity that very little is known about it. In the Acrasiales the regulation of the size of a fruiting body is relatively simple because its size is determined in advance, during aggregation of the amoebae. If aggregates do not break up, the number of cells in the fruiting body is the same or slightly higher than the number of amoebae forming the aggregate. The increase in the number of cells after aggregation depends on the degree of starvation when the amoebae came together (Sussman, R. R. and Sussman, M., 1960). The more the cells are starved prior to entering the aggregation, the fewer cells will divide later.

At a constant cell density, the size of an aggregate, and consequently of a fruiting body, depends on the area or territory covered by the aggregation. The numbers of aggregates and fruiting bodies per unit area are however identical only when aggregates do not fuse, break up into smaller pseudoplasmodia, or disintegrate. If the territory size in the larger *Dictyostelium* species were determined by the cyclic AMP production of an aggregate, the problem of size would be reduced to that of the scale of cyclic AMP production by the aggregates. Territory size would be dependent on population density if the increase in the size of an aggregate coincided with an increase in cyclic AMP production. This relationship exists only during early aggregation. When a larger number of cells enter the aggregate, the cyclic AMP production does not increase. In the case of *D. discoideum*, 500 amoebae will attract responding cells maximally. One consequence of this maximal output of attractant by a small group of cells is that territory size becomes independent of cell density. Over a large range of densities the size of the fruiting bodies increases proportionally to cell density and the area occupied by an aggregation stays constant.

As early as 1937, Arndt stated that territory size in *D. mucoroides* is nearly constant, regardless of the density of the bacterial layer and thus of the population density before aggregation. Bonner and Dodd (1962) provided support for this conclusion by quantitative data obtained in various species. They varied the population density by growing amoebae on an agar surface with various bacterial densities. The number of fruiting bodies per 0·102 cm^2 was counted. The aggregation territories of five species had different but constant sizes, even at high densities of the pre-aggregative amoebae. Aggregation on a glass surface of amoebae grown in liquid culture did not affect the constancy of the territory size. Changes in territory size can be induced by alteration of the environmental conditions. Both exposure to light and a decrease in humidity reduce territory size as do the addition of nutrients (Bonner and Dodd, 1962) or histidine (Bradley *et al.*, 1956) to the agar. The territory size is drastically reduced by the addition of charcoal (Bonner and Hoffman, 1963), which may remove gaseous inhibitors for the initiation of aggregation. Mineral oil is even more effective in reducing territory size. In *P. pallidum* the number of centres increases by a factor of 4 after exposure to charcoal and by a factor of 40 under oil. In *D. purpureum*, however, with or without charcoal, the number of centres per cm^2 is about the same and mineral oil does not seem to affect its territory size. Francis (1965), also working with *D. discoideum*, found a constancy in the number of aggregates over a range of population densities. The larger size of his territories (0·75 to 1·25 mm^2) may have been due to his use of a glass surface instead of agar; the attractant secreted by the amoebae can diffuse downward only on an agar surface.

The non-random distribution of centres is another indication that young aggregates suppress centre formation in their immediate neighbourhood (Bonner and Hoffman, 1963; Kahn, 1968). An inhibiting substance is thought to diffuse outward from young aggregates and prevent the formation of additional aggregates nearby (Shaffer, 1961). Francis (1965) suggested that the inhibition might be the result of a gradient of the attractant itself. The gaseous inhibitor and acrasin may both affect territory size.

The gaseous inhibitor is not species specific. All species tested secreted such an inhibitor, but not all species are sensitive to it (Bonner and Hoffman, 1963). Efforts are being made to identify the regulatory volatile substance which also determines the size of the fruiting body. Ammonia probably affects territory size (Feit, 1969; Lonski, 1973).

The number of centres in *D. purpureum* can also be increased by using amoebae harvested early in the aggregative phase. Reversal of the aggregative phase makes the centres formed by the reaggregating amoebae density dependent. At higher densities of this species more centres are formed per unit area (Bonner and Hoffman, 1963). The

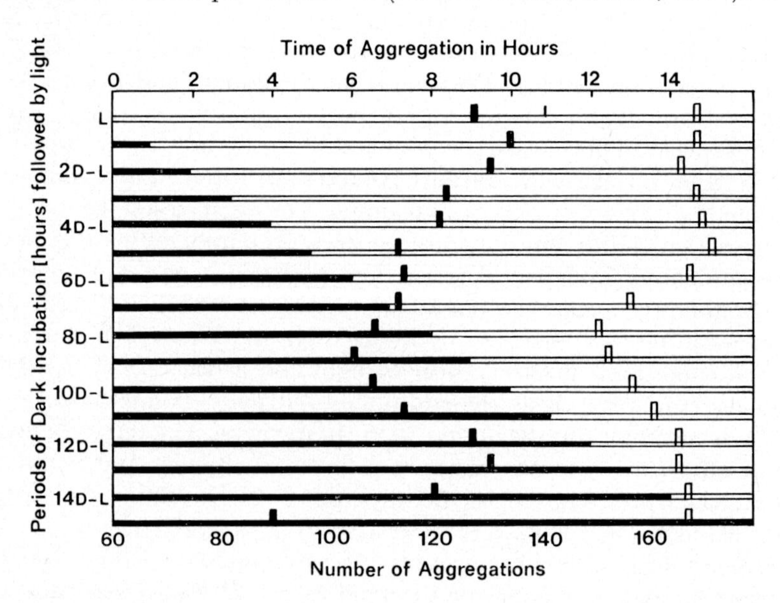

Fig. 11. The effect of length of dark period on the time of aggregation (□) and the number of aggregates (■). From Konijn and Raper (1966).

results obtained with centrifuged amoebae are not always consistent. The factors possibly responsible for the large variation in the number of centres per cm^2 include the developmental stage at which amoebae are harvested and the light conditions under which the cells are incubated.

Since light is involved in the determination of territory size in almost every species, the effect of light during the pre-aggregative phase on subsequent territory size was studied. When pre-grown amoebae were deposited on non-nutrient agar, more and smaller aggregates and fruiting bodies (Fig. 11) were formed in constant light than in darkness (Konijn and Raper, 1966). When the initial period of darkness was of optimal length for early aggregation in *D. discoideum*, the territory size was larger than under constant light but smaller than under constant darkness. Exposure of amoebae to light early in the aggregation phase sometimes resulted in the breaking up of aggregates into several smaller pseudoplasmodia. Therefore, for accurate measurement of the territory size it is not sufficient to count the number of sorocarps. Counts of aggregates must be made only if it is certain that they have not resulted from the break up of larger aggregates.

Other environmental conditions besides light, such as temperature, relative humidity and cell density, also affect territory size but the mechanism which determines the territory size and which is influenced by environmental conditions is not known. Environmental changes may influence acrasin production, secretion or both. Light may alter the sensitivity of responding amoebae, although this does not seem likely, because sensitive amoebae of *D. discoideum* in light and in darkness are attracted over the same distance by cyclic AMP secreted by bacteria (Konijn, 1969). When all aggregates are formed at the same time, competition plays a role. Amoebae at the edge of a territory may be exposed to acrasin from two competing centres. Whether centres are initiated simultaneously or appear during a period of several hours depends on environmental conditions.

Territory size is not independent of population density at very high densities; under these conditions territories will be small. The sticking of amoebae to each other before aggregation starts may affect the territory size at such high densities. A smaller territory size has also been observed at very low population densities (Sussman and Noël, 1952; Westra, personal communication). Possibly, at such low densities the aggregates do not secrete the attractant maximally, and the small aggregates consequently exert their influence over small areas.

The smaller aggregates formed in the light may result from reduced cyclic AMP secretion. The distance over which attraction by a centre is effective is less in the light than in darkness (Konijn and Raper, 1966). The shorter streams in light could also contribute to the reduced attraction in light, especially since cyclic AMP is secreted by streams as well as by the centre. The maximal binding activity of cyclic AMP does not play a role since binding in the light is similar to binding in darkness (Mato and Konijn, 1975).

If the territory size of a cellular slime mould depends on the maximal acrasin production of an aggregate and if one aggregate develops into one fruiting structure, the cellular slime moulds could be taken as model organisms to approach the problem of the determination of size of an organism. Preliminary results indicate that among the cyclic AMP sensitive species cyclic AMP production is correlated with the territory size and consequently with the size of the mature fruiting body (Westra, unpublished). It should be kept in mind that cell adhesion, sensitivity to acrasin, phosphodiesterase activity, length of streams, and inhibitors may also influence the relationship between size and acrasin concentration.

XII. Effect of the environment on aggregation

In the past, divergent results have been obtained on the aggregative behaviour of amoebae. Large variations in territory size under controlled environmental conditions point to a delicate balance between many factors governing territory sizes. The environmental conditions affect not only chemotaxis but also other phenomena contributing to cell aggregation. Changes in cell adhesion, permeability of plasma membranes, and metabolic processes inside the cell could all influence cell aggregation.

A. Light

That light causes small-sized fruiting bodies (Potts, 1902; Harper, 1932; Raper, 1940) has already been mentioned. The onset of aggregation is accelerated when amoebae are incubated in light (Raper, 1940; Shaffer, 1958). Pre-grown amoebae placed on plain agar and incubated under controlled conditions and known densities also aggregated earlier in constant light than in darkness, with the exception of *D. discoideum* (Konijn and Raper, 1965). Increase in length of dark periods delayed the onset of aggregation. One strain of *D. discoideum*, *Acr.* 12, behaved in a way similar to other species, but at a density of 200 cells per mm^2. All other strains of *D. discoideum* required an initial dark period followed by continuous light for early aggregation (Fig. 11). Light accelerates fructification in several newly described species of *Dictyostelium* and more and smaller fruiting structures are formed in light than in darkness (Raper and Fennell, 1967).

The light conditions under which amoebae are grown before harvesting also affects the time at which aggregation occurs. Amoebae grown in light, either on a solid medium or in shaken cultures, aggregate earlier than those grown in darkness (Konijn and Raper, 1965). Furthermore, the initial dark period optimal for aggregation is shorter for cells previously grown in light. In these studies the time of aggregation was defined as the moment when streams directed to a permanent centre became clearly visible.

To be optimally effective, the dark period in *D. discoideum* had to be at least 4 to 6 hours. The length of the initial dark period for early aggregation is shorter the higher the cell density. After some hours of incubation in darkness amoebae of *P. pallidum* will respond to one or two minutes of light (Kahn, 1964; Jones and Francis, 1972) by producing more aggregates. The greater sensitivity of *P. pallidum* to light is also indicated by the absence of aggregations in darkness (Kahn, 1964). Jones and Francis (1972), however, found several aggregates in drops of amoebae not exposed to light. The discrepancy may result from different growth conditions; Jones and Francis grew the amoebae in light, whereas Kahn grew his in darkness. *Acrasis rosea* requires both light and dark periods for fruiting, and does not fruit in either continuous light or continuous darkness (Olive and Stoianovitch, 1960; Reinhardt, 1968).

There is no evidence for photo-acceleration of the movement of amoebae (Francis, 1964) or phototaxis of single amoebae (Bonner and Whitfield, 1965) in cellular slime moulds, but pseudoplasmodia are phototactic. To identify the part of the spectrum responsible for the induction of aggregation in *P. pallidum*, Jones and Francis (1972) used the same intensity at each wave length and found two peaks of activity, a main peak at 475 nm and a minor peak at 675 nm. Kientzler and Zetsche (1972), who studied the effect of light on aggregation in the same species, observed peaks of activity at 460 nm and 600 nm. Reinhardt and Mancinelli (1968), who investigated the induction of aggregation in *A. rosea* also found a main peak at 450 nm. The phototaxis of pseudoplasmodia of *D. discoideum* (see Chapters 1 and 2) shows peaks at 430 and 560 nm (Francis, 1964; Poff *et al.*, 1973).

A major question is the extent to which light effects in the cellular slime moulds are tied to chemotaxis. It is conceivable that when amoebae are close to aggregation, illumination increases acrasin secretion and this in turn accelerates aggregation.

B. *Temperature*

Lower temperature delays the onset of aggregation (Potts, 1902; Raper, 1940). Another effect shared with darkness is an increase in the distance over which attraction will occur in the small population assay (Konijn, 1965). Perhaps under natural conditions both light and warmth accelerate the initiation of the social phase at sunrise. The optimal temperature for the growth of *D. discoideum*, 20–24°C (Raper 1940), does not coincide with the optimal temperature for maximal attraction of sensitive amoebae. In contrast to light, temperature affects amoebae only at the time of aggregation. The reduced chemotaxis at higher temperature is due to an effect on the aggregate and not to a reduced sensitivity of the responding cells (Konijn, 1969). Adhesion is reduced at low temperatures, which is possibly due to an inability to expand areas of mutual contact (Garrod and Born, 1971). Also a lower phosphodiesterase activity may contribute to the increased chemotaxis at low temperatures.

The interval from the onset to the completion of aggregation was twice as long at 13°C as at 22·5°C (Konijn, 1965). In a dish, 1 mg of cyclic AMP induces a chemotactic response in small populations of *D. discoideum* over a distance of 20 mm. At 6°C, the same quantity of cyclic AMP attracts amoebae over distances as large as 50 mm. Recently we have shown that maximal binding activity is higher at 15°C than at 22°C (Fig. 12). The increased binding at lower temperature will contribute to attraction over greater distances. The number of cyclic AMP receptors

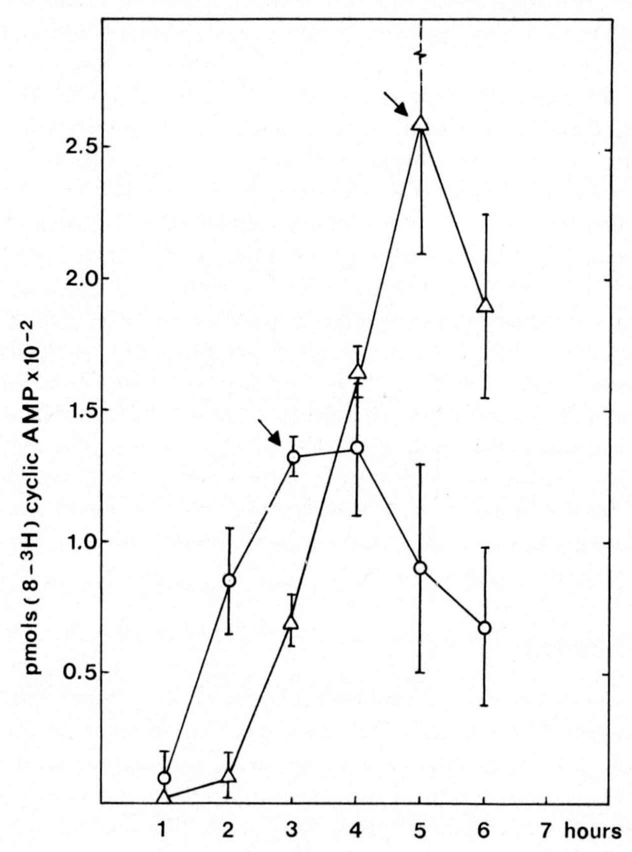

FIG. 12. Binding of [8-^{3}H] cyclic AMP in *D. mucoroides* at 22°C (o) and 15°C (△). ↘: onset of aggregation.

is independent of the temperature and 3 to 4 times higher in *D. discoideum* than in *D. mucoroides* (Mato and Konijn, 1975).

C. Humidity

The effect of humidity on chemotaxis is more difficult to quantity. Lower relative humidity results in earlier aggregation and smaller aggregates (Raper, 1940). Some species are very sensitive to the relative humidity with respect to aggregation and fructification. *D. polycephalum* requires a relative humidity of 98% for optimal fructification (Whittingham and Raper, 1957).

D. Cell density

Shaffer (1957) placed amoebae midway between acrasin-secreting sources differing in size. Amoebae between a thin aggregating stream and a thick stream or an aggregation centre moved approximately equally to both sides.

Studies in small populations of amoebae on a hydrophobic agar surface have shown that aggregates of *D. discoideum* attract sensitive cells over distances of 1·6 mm. The chemotactic response was independent of the number of cells in the attracting populations, which had a diameter of 0·6 mm and contained 400 to 5,000 amoebae (Konijn, 1968). A negative feedback mechanism might be responsible for the reduced output of attractant per cell at higher population densities. An increased phosphodiesterase activity at higher cell densities could keep the cyclic AMP level constant, but in that case it would be difficult to explain why there is not a clear effect of pH (which affects enzyme activity) on chemotaxis. A reduced acrasin secretion per cell (in *D. discoideum* a reduced cyclic AMP secretion) keeps the acrasin concentration and consequently the aggregation size within certain limits. The duration of aggregation was also independent of cell density. At all densities the interval between the beginning of aggregation and the moment at which almost all of the streams entered the centre was 3 to 3½ hours. A still lower population density resulted in a considerable delay in the onset of aggregation and therefore its effect on chemotaxis could not be measured.

Chemotaxis could be reduced by lowering the number of cells in populations with a diameter of 0·35 mm. The minimal number of amoebae required for maximal attraction was about 500 cells per population, and the chemotactic reaction of the amoebae in the responding drops seemed to be similar when 500 or more cells were present. At a density of 100 amoebae per responding drop, attraction was drastically reduced.

Responding populations of *P. pallidum* also reacted independently of the population density in the attracting drop over a wide range of densities (Konijn, 1973a). Francis (1965) found a higher acrasin production in larger aggregates of *P. pallidum*. These two observations are not contradictory, since Francis used a water-flow technique which meant that the chemotactic compounds were constantly washed away. In his experiments the acrasin concentration of *P. pallidum* could not reach the plateau at which acrasin production becomes independent of the size of the aggregate.

The effect of population density is not limited to the aggregative phase. During the interphase, the population density determines the time at which aggregation starts. At densities of 250 cells of *D. discoideum* per mm^2, there is an interval of 12 to 14 hours before the amoebae aggregate, and at high densities they enter the multicellular stage within a few hours.

Amoebae apparently sense the closeness of their neighbours. Although cells come into contact, especially shortly before aggregation, the earlier aggregation is more probably due to the exchange of substances (e.g. acrasin itself) responsible for shortening the prelude to aggregation.

E. *The pH level*

The pH of the environment does not seem to influence chemotaxis or cell aggregation significantly. Aggregation can occur from pH 4·0 to 8·0 (Hirschberg and Rusch, 1950), and the attraction of amoebae by bacteria in the small population assay is also independent of the pH within the range 5·0–8·0 (Konijn, unpublished).

F. *Inhibitors and stimulators*

The interval between the feeding stage and aggregation can be shortened in several species by the addition of activated charcoal. Inhibitory volatile substances are apparently absorbed by the charcoal. Histidine accelerates (Bradley *et al.*, 1956) and adenine retards aggregation (Krichevsky and Wright, 1961). The effect of substances that enhance or retard aggregation should be studied with respect to adenylate cyclase and phosphodiesterase activity. A specific inhibitor of phosphodiesterase has been discussed in section VII.

Because so many factors can interfere with chemotaxis and aggregation, experiments with cellular slime moulds should be carried out under closely controlled conditions and these should be clearly stated in all publications.

XIII. Initiation of aggregation

The initiation of aggregation is caused primarily by starvation (Potts, 1902; Arndt, 1937; Raper, 1940) and is influenced by environmental factors. Although aggregation has been studied in the greatest detail in *Dictyostelium*, its initiation has been observed most clearly in *Polysphondylium*. Certain cells of *P. violaceum* called founder cells become stationary, round up, and attract neighbouring cells (Shaffer, 1961). Shaffer postulated that the stimulated cells in their turn secrete acrasin to attract the more distant cells; increased adhesiveness would facilitate the multicellular stage. Founder cells appear spontaneously within a population of amoebae and may originate due to differences in the physiological age and environmental conditions. The founders do not differ morphologically and genetically from the other cells (Shaffer, 1961). Their ratio in the population is not constant and depends on cell density and environment. The sustained secretion of acrasin was demonstrated by Shaffer by lethal irradiation of the founder cell; the responding cells immediately stopped their oriented movement towards the dead founder cell. The same founder will attract the neighbouring cells even after dispersal has been induced repeatedly. Eventually, other cells may become founders. Founders kept isolated from other cells become responding cells. Light induces the formation of founders, whereas darkness delays their occurrence (Shaffer, 1961). Founders also occur in other species, e.g. *P. pallidum* (Francis, 1965) and *D. minutum* (Gerisch, 1964).

In *P. pallidum*, not all founders appear at the same time, and their occurrence is more abundant among older amoebae which have digested all their engulfed bacteria (Francis, 1965). The initiation of aggregation may therefore depend on differences in the concentration of attractants and the sensitivity of neighbouring cells that respond to the first founder secreting acrasin above threshold levels. Aggregation patterns resulting from differences in acrasin concentration would fit well with the theoretical model of Keller and Segel (1970), who assume the occurrence of instabilities in a field with a uniform distribution of continuously secreted acrasin. The model of Cohen and Robertson (1971a, b) depends on acrasin being secreted in pulses. Neighbouring cells attracted by the acrasin-secreting cell are prevented from initiating their own aggregates (Shaffer, 1961; Francis, 1965). Acrasin itself may exert this inhibiting effect (Francis, 1965). Founder cells of *D. minutum* do not inhibit the formation of other founders (Gerisch, 1965).

There is no clear evidence that cyclic AMP sensitive species use founder cells to initiate aggregation. Sussman and his associates (Ennis and Sussman, 1958; Sussman and Ennis, 1959) described in *D. discoideum* a special class of initiator cells occurring in a constant ratio of about 1:2,000, in the vegetative as well as in the preaggregative stage and easily detected because of their large size. The wide variation in the number of very large cells found by other authors and the frequent aggregation observed in the absence of initiator cells made the initiator cell hypothesis untenable (Gerisch, 1961; Konijn and Raper, 1961; Shaffer, 1962).

Perhaps the multicellular state in the larger *Dictyostelium* species is initiated by a small clump of cells but may be occasionally begun by a single cell. In a film of aggregation in a small population of *D. purpureum* I observed that one cell which happened to engulf a spore became the focal point of aggregation. The spore-containing cell moved more slowly than the others and may thus have been in a better position to attract neighbouring cells. The importance of retardation of cell movement was established by Samuel (1961), who measured the speed of movement of *D. mucoroides* amoebae in the vegetative and pre-aggregative phases. Other cells joining the attracting cell or cells can be expected to become synchronized in the secretion of pulses of cyclic AMP, synchronization being necessary to amplify the pulses which enable a small group of cells to attract all other amoebae in a drop measuring 0·3 mm^2. To be effective, synchronization of pulses of individual cells must take place within a few minutes. In other systems, such as the beat in isolated cardiac myocytes, synchrony occurs within minutes after the establishment of contacts (Mark and Strasser, 1966; DeHaan and Hirakow, 1972). In cellular slime moulds, too, synchronization of pulsations takes place shortly after contact. Using time-lapse cinematography, I followed the fusion of two aggregates in a small population of *D. discoideum*. Initially, the pulsations in the two aggregates were not entrained. The aggregates approached each other, and after they fused the one large aggregate emitted pulses at 5-minute intervals with pulsations occurring in phase.

The increased number of cells in an aggregate and their synchronous emission of cyclic AMP permits attraction of responding cells over distances amounting to more than a hundred times the diameter of the amoebae. Probably an amoeba moves towards an aggregate by measuring a spatial gradient of the attractant over its total length (Mato, Losada, Nanjundiah and Konijn, unpublished).

XIV. Conclusions

The cellular slime mould *Dictyostelium discoideum* responds to two distinct chemotactic agents, folic acid and cyclic AMP. Vegetative amoebae, actively feeding on bacteria, are particularly sensitive to folic acid and aggregating or pre-aggregative amoebae to cyclic AMP. Aggregates also emit cyclic AMP. It is therefore clear that cyclic AMP is the acrasin, the chemotactic agent responsible for aggregation, in *D. discoideum* whereas attraction of amoebae to bacteria is probably due largely to folic acid. The enzyme phosphodiesterase has a role in aggregation, destroying cyclic AMP and thus maintaining sharp gradients and preventing background "noise" due to excess attractant.

Cyclic AMP also attracts pre-aggregative amoebae of other *Dictyostelium* spp. with large fruiting bodies. However, there are other cellular slime moulds including *Polysphondylium pallidum* that do not respond to cyclic AMP by chemotaxis and for these other acrasins must be sought.

Other important questions concerning aggregation include the following:

(1) Does aggregation involve a relay system, i.e. do cells that respond to acrasin then secrete acrasin which attracts more distant cells, or does a centre, stream or group of amoebae itself attract the most distant cells?
(2) Is acrasin emitted continuously or in pulses?
(3) Is there a single founder cell for each aggregate or do many cells begin secreting acrasin at the same time?

It is probable that for each of these questions either answer is correct, depending on species and circumstances.

References

Aboud, M. and Burger, M. (1971). *Biochem. Biophys. Res. Commun.* **43,** 174–182.

Arndt, A. (1937). *Roux' Arch.* **136,** 681–744.

Barkley, D. S. (1969). *Science* **165,** 1133–1134.

Bonner, J. T. (1947). *J. Exp. Zool.* **106,** 1–26.

Bonner, J. T. (1949). *J. Exp. Zool.* **110,** 259–272.

Bonner, J. T. (1967). "The Cellular Slime Molds", 2nd edn. Princeton Univ. Press, Princeton.

Bonner, J. T. (1970). *Proc. Nat. Acad. Sci., U.S.A.*, **65,** 110–113.

Bonner, J. T. (1974). *In* "Humoral Control in Growth and Differentiation" (J. Lobue and A. S. Gordon, Ed.), Vol. 2, pp. 81–98. Academic Press, New York.

Bonner, J. T. and Dodd, M. R. (1962). *Develop. Biol.* **5,** 344–361.

Bonner, J. T. and Hoffman, M. E. (1963). *J. Embryol. Exp. Morph.* **11,** 571–589.

Bonner, J. T. and Whitfield, F. E. (1965). *Biol. Bull.* **128,** 51–57.

Bonner, J. T., Kelso, A. P. and Gillmor, R. G. (1966). *Biol. Bull.* **130,** 28–42.

Bonner, J. T., Barkley, D. S., Hall, E. M., Konijn, T. M., Mason, J. W., O'Keefe, G. and Wolfe, P. B. (1969). *Develop. Biol.* **20,** 72–87.

Bonner, J. T., Hall, E. M., Sachsenmaier, W. and Walker, B. K. (1970). *J. Bacteriol.* **102,** 682–687.

Bonner, J. T., Hirshfield, M. F. and Hall, E. M. (1971). *Exp. Cell Res.* **68,** 61–64.

Bonner, J. T., Sieja, T. W. and Hall, E. M. (1971). *J. Embryol. Exp. Morph.* **25,** 457–465.

Bonner, J. T., Hall, E. M., Noller, S., Oleson, F. B. and Roberts, A. B. (1972). *Develop. Biol.* **29,** 402–409.

Bradley, S. G., Sussman, M. and Ennis, H. L. (1956). *J. Protozool.* **3,** 33–38.

Brooker, G., Thomas, L. J. and Appleman, M. M. (1968). *Biochem.* **7,** 4177–4181.

Butcher, R. W. and Sutherland, E. W. (1962). *J. Biol. Chem.* **237,** 1244–1250.

CHANG, Y.-Y. (1968). *Science*. **160,** 57–59.
CHASSY, B. M., LOVE, L. L. and KRICHEVSKY, M. I. (1969). *Fed. Proc.* **28,** 842.
CHASSY, B. M. (1972). *Science* **175,** 1016–1018.
CHI, Y.-Y. and FRANCIS, D. (1971). *J. Cell. Physiol.* **77,** 169–174.
COHEN, M. H. and ROBERTSON, A. (1971a). *J. Theor. Biol.* **31,** 101–118.
COHEN, M. H. and ROBERTSON, A. (1971b). *J. Theor. Biol.* **31,** 119–130.
COHEN, M. H. and ROBERTSON, A. (1972). *In* "Celli Dfferentiation" (R. Harris, P. Allin and D. Viza, eds.), pp. 35–45. Munksgaard, Copenhagen.
D'ARMIENTO, M., JOHNSON, G. S. and PASTAN, I. (1972). *Proc. Nat. Acad. Sci. U.S.A.*, **69,** 459–462.
DEHAAN, R. L. and HIRAKOW, R. (1972). *Exp. Cell Res.* **70,** 214–220.
DURSTON, A. J. (1974). *Develop. Biol.* **38,** 308–319.
ELLOUZ, R. and LENFANT, M. (1971). *Eur. J. Biochem.* **23,** 544–550.
ENNIS, H. L. and SUSSMAN, M. (1958). *Proc. Nat. Acad. Sci. U.S.A.* **44,** 401–411.
FEIT, I. (1969). *Ph.D. Thesis*, Princeton University.
FRANCIS, D. W. (1964). *J. Cell. Comp. Physiol.* **64,** 131–138.
FRANCIS, D. W. (1965). *Develop. Biol.* **12,** 329–346.
GARROD, D. R. and BORN, G. V. R. (1971). *J. Cell Sci.* **8,** 751–765.
GARROD, D. R. and MALKINSON, A. M. (1973). *Exp. Cell Res.* **81,** 492–495.
GERISCH, G. (1961). *Develop. Biol.* **3,** 685–724.
GERISCH, G. (1964). *Arch. Entwicklungsmech. Organismen* **155,** 342–357.
GERISCH, G. (1965). *Arch. Entwicklungsmech. Organismen* **156,** 127–144.
GERISCH, G. (1966). *Arch. Entwicklungsmech. Organismen* **157,** 174–189.
GERISCH, G. (1968). *Current Topics in Develop. Biol.* **3,** 157–197.
GERISCH, G., NORMANN, I. and BEUG, H. (1966). *Naturwiss.* **23,** 618–619.
GERISCH, G., MALCHOW, D., RIEDEL, V., MÜLLER, E. and EVERY, M. (1972). *Nature New Biol.* **235,** 90–92.
GERISCH, G. and HESS, B. (1974). *Proc. Nat. Acad. Sci. U.S.A.*, **71,** 2118–2122.
GILLEBRAND, I. M. (1970). *Biochem. J.* **120,** 20P–21P.
GINGELL, D. and GARROD, D. R. (1969). *Nature* **221,** 192–193.
GOIDL, E. A., CHASSY, B. M., LOVE, L. L. and KRICHEVSKY, M. I. (1972). *Proc. Nat. Acad. Sci. U.S.A.* **69,** 1128–1130.
GREENGARD, P., HAYAISHI, O. and COLOWICK, S. P. (1969). *Fed. Proc.* **28,** 467.
GREGG, J. H. and NESOM, M. G. (1973). *Proc. Nat. Acad. Sci. U.S.A.* **70,** 1630–1633.
HARPER, R. A. (1932). *Bull. Torrey Bot. Club* **59,** 49–84.
HEFTMANN, E., WRIGHT, B. E. and LIDDEL, G. U. (1960). *Arch. Biochem. Biophys.* **91,** 266–270.
HERRMANN-ERLEE, M. P. M. and KONIJN, T. M. (1970). *Nature* **227,** 177–178.
HIRSCHBERG, E. and RUSCH, H. P. (1950). *J. Cell. Comp. Physiol.* **36,** 105–113.
HOSTAK, M. B. and RAPER, K. B. (1960). *Bacteriol. Proc.* **60,** 58–59.
JONES, W. R. and FRANCIS, D. (1972). *Biol. Bull.* **142,** 461–469.
KAHN, A. J. (1964). *Biol. Bull.* **127,** 85–96.
KAHN, A. J. (1968). *Develop. Biol.* **18,** 149–162.
KAKIUCHI, S. (1971). *Biochem. Biophys. Res. Commun.* **42,** 968–974.
KELLER, E. F. and SEGEL, L. A. (1970). *J. Theor. Biol.* **26,** 399–416.
KIENTZLER, M. and ZETSCHE, K. (1972). *Naturwiss.* **59,** 40.
KONIJN, T. M. (1965). *Develop. Biol.* **12,** 487–497.

Konijn, T. M. (1968). *Biol. Bull.* **134,** 298–304.
Konijn, T. M. (1969). *J. Bacteriol.* **99,** 503–509.
Konijn, T. M. (1970). *Experientia* **26,** 367–369.
Konijn, T. M. (1972a). *In* "Advances in Cyclic Nucleotide Research" (P. Greengard, G. A. Robison and R. Paoletti, eds.), Vol. I, pp. 17–31. Raven Press, New York.
Konijn, T. M. (1972b). *Acta Protozool.* **11,** 137–144.
Konijn, T. M. (1972c). *In* "Institutes of the Royal Neth. Acad. of Arts and Sciences", p. 47.
Konijn, T. M. (1973a). *In* "Behaviour of Micro-organisms" (A. Pérez-Miravete, ed.), pp. 48–61. Plenum Press, London.
Konijn, T. M. (1973b). *FEBS Letters* **34,** 263–266.
Konijn, T. M. (1974). In "Chemotaxis: Its Biology and Biochemistry" (E. Sorkin, ed.) pp. 96–110. Karger, Basel.
Konijn, T. M. and Raper, K. B. (1961). *Develop. Biol.* **3,** 725–756.
Konijn, T. M. and Raper, K. B. (1965). *Biol. Bull.* **128,** 392–400.
Konijn, T. M. and Raper, K. B. (1966). *Biol. Bull.* **131,** 446–456.
Konijn, T. M., Van De Meene, J. G. C., Bonner, J. T. and Barkley, D. S. (1967). *Proc. Nat. Acad. Sci., U.S.A.* **58,** 1152–1154.
Konijn, T. M., Barkley, D. S., Chang, Y. Y. and Bonner, J. T. (1968). *Amer. Natur.* **102,** 225–234.
Konijn, T. M., Van De Meene, J. G. C., Chang, Y. Y., Barkley, D. S. and Bonner, J. T. (1969a). *J. Bacteriol.* **99,** 510–512.
Konijn, T. M., Chang, Y. Y. and Bonner, J. T. (1969b). *Nature,* **224,** 1211–1212.
Konijn, T. M. and Koevenig, J. L. (1971). *Mycologia* **63,** 901–906.
Konijn, T. M. and Jastorff, B. (1973). *Biochem. Biophys. Acta* **304,** 774–780.
Krichevsky, M. I. and Wright, B. (1961). *Bacteriol. Proc.*, p. 86.
Leach, C. K., Ashworth, J. M. and Garrod, D. R. (1973). *J. Embryol. Exp. Morph.* **29,** 647–661.
Lonski, J. (1973). Ph. D. Thesis, Princeton University.
McManus, J. P. and Whitfield, J. F. (1969). *Exp. Cell Res.* **58,** 188–190.
Maeda, Y. (1970). *Develop. Growth Differ.* **12,** 217–228.
Makman, M. H. and Sutherland, E. W. (1965). *J. Biol. Chem.* **240,** 1309–1314.
Malchow, D., Nägele, B., Schwarz, H. and Gerisch, G. (1972). *Eur. J. Biochem.* **28,** 136–142.
Malchow, D., Fuchila, J. and Jastorff, B. (1973). *FEBS Letters* **34,** 5–9.
Malchow, D. and Gerisch, G. (1973). *Biochem. Biophys. Res. Commun.* **55,** 200–204.
Malkinson, A. M. and Ashworth, J. M. (1972). *Biochem. J.* **127,** 611–612.
Malkinson, A. M. and Ashworth, J. M. (1973). *Biochem. J.* **134,** 311–319.
Malkinson, A. M., Kwasniak, J. and Ashworth, J. M. (1973). *Biochem. J.* **133,** 601–603.
Mark, G. E. and Strasser, F. F. (1966). *Exp. Cell Res.* **44,** 217–233.
Mason, J. W., Rasmussen, H. and Dibella, F. (1971). *Exp. Cell Res.* **67,** 156–160.
Mato, J. M. and Konijn, T. M. (1975). *Biochim. Biophys Acta,* **385,** 173–179.

MOENS, P. B. and KONIJN, T. M. (1974). *FEBS Letters* **45,** 44–46.
MURRAY, A. W., SPISZMAN, M. and ATKINSON, D. E. (1971). *Science* **171,** 496–498.
NESTLE, M. and SUSSMAN, M. (1972). *Develop. Biol.* **28,** 545–554.
OLIVE, E. W. (1902). *Proc. Boston Soc. Nat. Hist.* **30,** 451–513.
OLIVE, L. S. and STOIANOVITCH, C. (1960). *Bull. Torrey Bot. Club* **87,** 1–20.
PAN, P., HALL, E. M. and BONNER, J. T. (1972). *Nature New Biol.* **237,** 181–182.
PAN, P., BONNER, J. T., WEDNER, H. J. and PARKER, C. W. (1974). *Proc. Nat. Acad. Sci. U.S.A.* **71,** 1623–1625.
PANNBACKER, R. G. and BRAVARD, L. J. (1970). *Bacteriol. Proc.* **70,** 23.
PASTAN, I. and PERLMAN, R. (1970). *Science* **169,** 339–344.
POFF, K. L., LOOMIS, W. F. and BUTLER, W. L. (1974) *J. Biol. Chem.* **249,** 2164–2167.
POTTS, G. (1902). *Flora* **91,** 281–347.
RAPER, K. B. (1940). *J. Elisha Mitchell Sci. Soc.* **56,** 241–282.
RAPER, K. B. (1941). *Mycologia* **33,** 633–649.
RAPER, K. B. (1960). *Proc. Amer. Phil. Soc.* **104,** 579–604.
RAPER, K. B. (1973). *In* "The Fungi" (G. C. Ainsworth, F. K. Sparrow and A. S. Sussman, eds.), Vol. 4B, pp. 9–36. Academic Press, New York.
RAPER, K. B. and THOM, C. (1941). *Amer. J. Bot.* **28,** 69–78.
RAPER, K. B. and FENNELL, D. I. (1967). *Amer. J. Bot.* **54,** 515–528.
RASMUSSEN, H. (1970). *Science* **170,** 404–412.
RASMUSSEN, H. and TENENHOUSE, A. (1968). *Proc. Nat. Acad. Sci. U.S.A.* **59,** 1364–1370.
REINHARDT, D. J. (1968). *Amer. J. Bot.* **55,** 77–86.
REINHARDT, D. J. and MANCINELLI, A. L. (1968). *Develop. Biol.* **18,** 30–41.
RIEDEL, V. and GERISCH, G. (1971). *Biochem. Biophys. Res. Commun.* **42,** 119–123.
RIEDEL, V., GERISCH, G., MÜLLER, E. and BEUG, H. (1973). *J. Mol. Biol.* **74,** 573–585.
ROBERTSON, A., DRAGE, D. J. and COHEN, M. H. (1972). *Science* **175,** 333–335.
ROBISON, G. A., BUTCHER, R. W. and SUTHERLAND, E. W. (1971). *Cyclic AMP.* Academic Press, New York.
ROSSOMANDO, E. F. and SUSSMAN, M. (1972). *Biochem. Biophys. Res. Commun.* **47,** 604–610.
ROSSOMANDO, E. F. and SUSSMAN, M. (1973). *Proc. Nat. Acad. Sci. U.S.A.* **70,** 1254–1257.
RUNYON, E. H. (1942). *Collecting Net.* **17,** 88.
SAMUEL, E. W. (1961). *Develop. Biol.* **3,** 317–336.
SHAFFER, B. M. (1956a). *J. Exp. Biol.* **33,** 645–657.
SHAFFER, B. M. (1956b). *Science* **123,** 1172–1173.
SHAFFER, B. M. (1957). *Amer. Natur.* **91,** 19–35.
SHAFFER, B. M. (1958). *Quart. J. Microsc. Sci.* **99,** 103–121.
SHAFFER, B. M. (1961). *J. Exp. Biol.* **38,** 833–849.
SHAFFER, B. M. (1962). *Advan. Morphogenesis* **2,** 109–182.
SHIMOYAMA, M., KAWAI, M., TANIGAWA, Y., UEDA, I., SAKAMOTO, M., HAGIWARA, K., YAMASHITA, Y. and SAKAKIBARA, E. (1972). *Biochem. Biophys. Res. Commun.* **47,** 59–65.

SONNEBORNE, D. R., WHITE, G. H. and SUSSMAN, M. (1963). *Develop. Biol.* **7,** 79–93.

SPANGLER, R. A. and SNELL, F. M. (1961). *Nature* **191,** 457–458.

SUSSMAN, M. (1955). *J. Gen. Microbiol.* **13,** 295–309.

SUSSMAN, M. and NOËL, E. (1952). *Biol. Bull.* **103,** 259–268.

SUSSMAN, M., LEE, F. and KERR, N. S. (1956). *Science* **123,** 1171–1172.

SUSSMAN, R. R., SUSSMAN, M. and FU, F. L. (1958). *Bacteriol. Proc.*, p. 32.

SUSSMAN, R. R. and SUSSMAN, M. (1960). *J. Gen. Microbiol.* **23,** 287–293.

SUTHERLAND, E. W. and RALL, T. W. (1958). *J. Biol. Chem.* **232,** 1077–1091.

SUTHERLAND, E. W., RALL, T. W. and MENON, T. (1962). *J. Biol. Chem.* **237,** 1220–1227.

TAKEUCHI, I. (1969). *In* "Nucleic Acid Metabolism, Cell Differentiation and Cancer Growth" (E. V. Cowdry and S. Seno, eds.), pp. 297–304. Pergamon Press, Oxford and New York.

TAKEUCHI, I. and TAZAWA, M. (1955). *Cytologia* **20,** 157–165.

TAO, M. and LIPMANN, F. (1969). *Proc. Nat. Acad. Sci. U.S.A.* **63,** 86–92.

UNO, I. and ISHIKAWA, T. (1973). *J. Bacteriol.* **113,** 1240–1248.

VEERDONK, F. C. G. VAN DE and KONIJN, T. M. (1970). *Acta Endocrinol.* **64,** 364–376.

WEINSTEIN, B. I. and KORITZ, S. B. (1973). *Develop. Biol.* **34,** 159–162.

WHITTINGHAM, W. F. and RAPER, K. B. (1957). *Amer. J. Bot.* **44,** 619–627.

WIJK, R. VAN and KONIJN, T. M. (1971). *FEBS Letters,* **13,** 184–186.

WRIGHT, B. E. and ANDERSON, M. L. (1958). Symp. on "The Chemical Basis of Development" (W. O. McElroy and B. Glass, eds.), pp. 296–313. Johns Hopkins Press, Baltimore.

YOKOTA, T. and GOTS, J. S. (1970). *J. Bacteriol.* **103,** 513–516.

Chapter 5

Chemotaxis and chemotropism in fungi and algae

GRAHAM W. GOODAY

Department of Biochemistry, University of Aberdeen, Aberdeen, Scotland

I. Introduction

This chapter examines the oriented responses of growth and movement of fungal hyphae and of fungal and algal motile cells to gradients of chemicals in their environments. Four types of chemotropism and chemotaxis will be considered, although in some cases these classifications are not mutually exclusive: (1) Responses to extrinsic chemicals from possible nutrient sources; (2) Responses to extrinsic chemicals from possible hosts; (3) Responses to the presence of other cells of the same species involving oriented vegetative growth mediated either directly by specific intrinsic factors or indirectly by the effect of a neighbouring cell on the chemical environment; (4) Responses involving very specific sexual attractions. In all these cases there is evidence for true taxes and tropisms by oriented movement or growth in response to a concentration gradient.

II. Tropism and taxis towards nutrients

It is often assumed that fungal hyphae grow chemotropically towards nutrients. For most fungi, evidence for such a phenomenon is poor or non-existent. Stadler (1952) summarized many of his experiments with *Rhizopus nigricans* and other fungi thus: "Outside the staling reaction, no true tropic response has been detected for a large number of compounds and mixtures tested". Obviously a hypha growing in a local nutrient-rich substrate will proliferate and branch more than a neighbouring sibling hypha in a nutrient-poor substrate, but this involves no attraction to the nutrients.

Autotropism has been misinterpreted as chemotropism to nutrients, for example by Miyoshi (1894), who used a variety of techniques to investigate tropisms of several fungi towards a wide range of ammonium and other salts, sugars, peptone, and other nutrients. His techniques, such as using perforated mica sheets to separate two cultures, have formed the basis for nearly all subsequent chemotropic experiments. He also vacuum-infiltrated leaves with solutions to be tested, spread spores over the epidermis, and observed if the germ tubes grew towards the stomata. He reported positive chemotropism for all the fungi towards a wide range of nutrients. Clark (1902) used these techniques to investigate whether copper fungicides could elicit chemotropism and realized that Miyoshi had been observing the negative autotropism of germ-tubes away from their neighbours, and not positive chemotropism. Clark found that copper ions had no effect on the direction of growth of hyphae. Fulton (1906), who gave a comprehensive review of earlier work, confirmed Clark's results and as a result of extensive tests with many compounds concluded that the fungi tested had "no definite sensibility to nutrient substances", and that the tropisms observed were due to negative autotropism or perhaps sometimes to positive tropism to water vapour. Graves (1916) reported a slight tropic effect towards nutrients, especially turnip juice, that was much less than the negative autotropism and so was very easily overlooked. Stadler (1952, 1953) reinvestigated Graves' observations, and studied the effect of different concentrations of spores on the 'attraction' of germ-tubes to turnip juice and other nutrients. He concluded that no true chemotropism is involved, but that such nutrients interfere with autotropism by interactions with the staling substances or their formation or action. Robinson (1973) interpreted this apparent chemotropism to turnip juice as a positive chemotropism to oxygen gradients. He suggested that cells exposed to the nutrients in the turnip juice would respire faster, and give rise to an oxygen gradient towards the turnip juice, in that direction. Thus there is no clear evidence for chemotropism towards nutrients (except perhaps oxygen, if this is regarded as a nutrient) in the Zygomycetes, Ascomycetes or Basidiomycetes. However, chemotropism of hyphae and chemotaxis of zoospores to nutrients does occur in the aquatic Phycomycetes, which are considered phylogenetically distinct from other fungi.

Fischer and Werner (1955, 1958) investigated chemotropism and chemotaxis in a number of species of water-moulds in the Oomycetes, particularly *Saprolegnia* and *Achlya*. They placed blocks of agar containing nutrients near vegetative hyphae of *Saprolegnia ferax* and *Saprolegnia mixta* and observed strong chemotropism towards agar containing a mixture of amino acids at 2·5 mg/ml. If the blocks were placed on older vegetative mycelium the hyphal tips grew around and back towards the source of amino acids. The chemotropically attracted hyphae showed increased branching, and all the branches also grew towards the stimulus, as did germ tubes of germinating spores. These authors also found strong chemotropism of vegetative hyphae and antheridial branches of *Achlya polyandra* towards agar containing 20 mg/ml casein hydrolysate. This stimulus proved greater than the sexual chemotropism of these antheridial branches towards neighbouring oogonia that would otherwise have occurred and which is probably mediated by antheridiol (see below). Barksdale (1969) observed that antheridiol-stimulated branches of *Achlya ambisexualis* stay as antheridial hyphae in the presence of low nutrient levels, but become vegetative hyphae at higher nutrient concentrations. Fischer and Werner (1955) also investigated the responses of a wide range of other fungi growing on 0·2% malt extract to agar blocks containing 5 mg/ml casein hydrolysate. Strong positive chemotropism was shown by eighteen species of Oomycetes, less strong chemotropism by a further nineteen species of Oomycetes, and none by Zygomycetes, Ascomycetes and Basidiomycetes. Individual amino acids were not effective. The simplest active mixture was glutamic acid, leucine and cysteine, but mixtures of at least five L-amino acids were required for a strong chemotropic response. Fischer and Werner (1958) then studied the chemotaxis of the zoospores of Oomycetes to their natural substrates of animal and plant debris in water, described by Pfeffer (1884) and Müller (1911). Using *S. mixta* and *S. ferax* they showed that a range of salts would attract the zoospores. Potassium chloride was the most efficient, followed by other alkali metal chlorides and calcium and magnesium chlorides. Proteins were inactive, but a mixture of amino acids and salts gave good attraction and would completely simulate natural substrates.

Carlile and Machlis (1965b) and Machlis (1969a, b) investigated the chemotaxis of zoospores of the Chytridiomycete water moulds, *Allomyces macrogynous* and *Allomyces arbuscula*. The bio-assay involved counting the number of zoospores that settled on a cellophane membrane separating them from the test solution, and had the advantage over capillary tube methods that killed or partially immobilized cells were not counted. Active solutions, such as casein hydrolysate, attracted zoospores to the membrane within a few minutes and continued to do so for several hours. The number of zoospores attached to the membrane after 90 minutes was counted. Haploid zoospores, diploid zoospores and zygotes were attracted by casein hydrolysate, but not by sugars. The most active amino acids in the casein hydrolysate were leucine and lysine. Only L-amino acids were effective, and leucine and lysine acted synergistically, the optimum concentrations being 5×10^{-4} M. Attraction was still detectable with 1×10^{-5} M solutions. The addition of L-proline increased the response. Noting that the final response to these amino acids is the settling of the zoospores, Dill and Fuller (1971) and Olson and Fuller (1971) tested the effect of relatively high concentrations of the amino-acids on swimming zoospores of *Allomyces neo-moniliformis*. As predicted, the zoospores very quickly stopped swimming, and their flagella were found to be immobilized, sticking out as straight rods. Viability was not affected and synchronous development of the germlings occurred on removal of the amino acids. A mixture of 0·05 M L-leucine and L-lysine gave this characteristic response. Although other amino acids including D-leucine and D-lysine gave a similar response, longer exposure times and higher concentrations were required, and Dill and Fuller suggested that the chemotaxis and the immobilization could be related phenomena.

Zoospores and germ-tubes of some plant pathogenic fungi show chemotaxis and chemotropism towards a range of chemicals, probably by the same mechanisms as those of the saprophytic fungi described here, and are discussed in detail in the following section.

III. Attraction of parasites to their hosts

It has long been suggested that parasitic fungi find their hosts by chemotaxis or chemotropism, but only recently has there been any firm evidence for this. Some early reports can be reinterpreted as autotropism within the inoculum. For example, Massee (1905) proposed that germ-tubes of parasitic fungi grew directly towards susceptible host plants, attracted by specific chemicals, but he was using the techniques of Miyoshi (1894) which as discussed earlier show autotropism of germinating spores. There is, however, no doubt that the biflagellate zoospores of *Phytophthora*, *Pythium* and related species are attracted to the roots of their hosts, to the roots of non-host plants, and to root exudates and extracts in capillary tubes (Hickman, 1970; Hickman and Ho, 1966). The zoospores of these fungi are important infective agents from plant to plant in the soil, and their chemotactic attraction to roots of potential hosts could be a major factor in the spread of disease. The phenomenon was reported by Goode (1956) who observed that zoospores of *Phytophthora fragariae* gathered and encysted just behind the root tips of strawberry (the host plant) and of several other plants that were not hosts. On germination, all of the germ-tubes from the cysts grew towards the roots, where they could penetrate and infect. Some zoospores also encysted among the root hairs, but on germination their germ-tubes grew at random, and Goode interpreted the location of these zoospores as being the result of mechanical trapping rather than chemotaxis. Similar observations have been made for many other plant pathogenic Phycomycete species, and Hickman (1970) listed nineteen species in which a rapid accumulation of zoospores behind root tips has been observed *in vitro*. The exact response seems to vary with the species of fungi and higher plants used and the experimental conditions, and Hickman illustrated four patterns of zoospore accumulation—the typical pattern of zoospores at the elongation region just behind the root-tip; zoospores forming a sheath around the root, but concentrated just behind the tip; zoospores accumulating only at the sites of wounds in the roots; random encystment of zoospores. Royle and Hickman (1964a) showed that zoospores of *Pythium aphanidermatum* are very strongly attracted to the cut ends of excised root segments of pea (the host) and of many other plants. In some cases a transitory repulsion was seen for about 30 seconds after immersing the cut root into a zoospore suspension, after which the zoospores accumulated. When thicker sections of older roots were used the zoospores accumulated at the stele of the cut surface. The accumulation of zoospores at intact or cut roots can be correlated with sites of maximum exudation. Royle and Hickman (1964a) attempted to quantify the attraction to capillary tubes containing exudates and extracts by measuring the angle between the track of a zoospore as shown on a darkfield long-exposure photograph and the direction of the capillary mouth. They obtained figures of $15 \pm 10°$ (42 measurements) for exudate and $41 \pm 26°$ for water agar. Visual observation confirmed that the former zoospores were swimming towards the capillary, while the latter zoospores swam at random. The swimming speeds of the zoospores were $0{\cdot}66 \pm 0{\cdot}19$ units and $1{\cdot}07 \pm 0{\cdot}16$ units for exudate and water agar respectively. Thus in the vicinity of the exudate the attracted zoospores swam more slowly and so were trapped.

Mehrotra (1970) showed that attraction and encystment can occur in soil. He grew several host or non-host plants in cellulose acetate tubes (of Millipore membranes with 1 μm pores) at an angle so that roots grew adpressed to the membrane. The tubes were surrounded by non-sterile soil in an outer cylinder. The entire apparatus was immersed for three to four hours in zoospores suspensions of *Phytophthora drechsleri* or *Phytophthora megasperma* var. *sojae*, or a suspension of zoospores was added to the outer soil. The cellulose acetate membrane was dissected out and stained. Thousands of zoospores had accumulated in the region exactly corresponding to the root tip on the other side of the membrane. This was followed by more direct experiments in which zoospore suspensions were added to the soil of plant pots containing seedlings. The roots were later removed and stained with a fluorescent Calcofluor dye that sensitively detects zoospores, cysts and germ-tubes. After only four hours the zoospores had clearly accumulated on the roots, and later the germ tubes of their cysts were growing directly towards the roots.

Thus zoospores of these fungi will accumulate at the roots of higher plants by chemotaxis, and the germ-tubes of their cysts will grow towards the roots by chemotropism. These phenomena seem not to be host-specific. In only one case is there evidence suggesting a specific attraction to a host plant. Zentmeyer (1961, 1966, 1970) showed that zoospores of *Phytophthora cinnamomi* are attracted by roots of its host, avocado, particularly by the more susceptible varieties, but not by roots of non-host plants even though their root exudates contain glutamic and aspartic acids which are strongly attractive to the zoospores (Table I).

TABLE I. Attraction of Zoospores of *Phytophthora cinnamomi* to roots of a host plant (Avocado), and a non-host plant (Citrus)[a]

Distance from root (mm)	Average number of zoospores/0·5 mm square	
	Avocado	Citrus
0–0·5	34·0	0·6
0·5–1·0	14·7	1·4
1·0–1·5	11·1	0·9
1·5–2·0	8·7	1·3
2·0–2·5	5·0	0·9
2·5–3·0	4·2	1·3

[a] From Zentmeyer (1961).

Zentmeyer suggested that the latter exudates might contain inhibitors of chemotaxis not present in exudates from the avocado, or that the ratio between metabolities in the exudates is a factor of importance in determining whether chemotaxis occurs.

The identification of the attractants in root exudates would at first sight seem amenable to modern techniques of analysis, but has presented major problems. Two approaches have been used, the analysis of root exudates known to attract, and chemotactic tests on pure chemicals. However, a very wide range of chemicals can induce zoospore accumulation, including amino acids, carbohydrates and inorganic acids (Hickman, 1970). Another complication is the electrotaxis (response to electric currents) and rheotaxis (response to localized water currents) exhibited by zoospores (Katsura and Miyata, 1971). Indeed, Troutman and Wills (1964) suggested that the accumulation of zoospores of *Phytophthora parasitica* var. *nicotianae* on tobacco roots was an electrotactic response to the negative charge on the root at the region of elongation and not chemotaxis, but the electrotactic response of zoospores has been shown to depend very much on other ions and non-ionic components present in the medium (Katsura *et al.*, 1966).

The chemotaxis of zoospores to single chemicals can be very marked, but is always less than that shown to root exudates (Rai and Strobel,

TABLE II. Accumulations of zoospores of two isolates of *Pythium aphanidermatum* in response to exudates and mixtures in capillary tubes.

	Number of zoospores after 25 min[a]	Accumulation ratio[b]
Pea root extract		
4 ml/ml roots	257	—
80 ml/ml roots	16	—
Pea root exudate		
2 ml/ml roots	167	—
10 plants/ml	—	18·4 ± 3·3
Casein hydrolysate (1%)	4	3·9 ± 0·6
Ammonium glutamate (0·75%)	121	4·3 ± 0·9
Glucose, fructose, sucrose (1%)	6	—
Casein hydrolysate (1%) + 3 sugars (1%)	123	—
Complete amino acids and sugars from exudate (10 plants/ml)	—	11·0 ± 0·2
Water agar	0·5	1·1 ± 0·2

[a] From Royle and Hickman (1964b).
[b] From Chang-Ho and Hickman (1970). The ratio between the number of zoospores adjacent to the source of the extract and the number randomly dispersed.

1966). Mixtures of chemicals can elicit much stronger responses than the separate chemicals, Royle and Hickman (1964b) showed that zoospores of *P. aphanidermatum* are only weakly attracted to amino acids and sugars but mixtures of these compounds gave a greatly increased response of the same order as that of root exudates (Table II). These authors investigated the effects of a wide range of chemicals and mixtures on a variety of responses shown by zoospores: 'disorientation', of the typical gliding corkscrew swimming; 'attraction' (true chemotaxis), towards the source of the chemical; 'trapping', when zoospores are unable to escape from a diffusion shell around the capillary tube mouth (this response is not necessarily preceded by chemotaxis and could be due to toxic or narcotic effects); 'milling', of trapped zoospores around the mouth of the tube; 'arc-attraction', which results in a concentration of zoospores at a distance to the mouth of the tube, repelled from a high concentration or unfavourable pH; 'repulsion'; 'indifference'; and encystment. Glutamic acid (0·75%), buffered by a weak base, was the only single compound to elicit the chemotropism and encystment (Table II). Closely related compounds, such as glutamine, α-keto-glutaric acid and glutaric acid, did not cause attraction, nor did glutamic acid buffered by a strong base. Among other reports of chemotactically active substances are sugars and casein hydrolysate for *P. parasitica* var. *nicotianae* (Dukes and Apple (1961), several amino acids including glutamate for several *Phytophthora* species (Zentmeyer, 1966; Khew and Zentmeyer, 1973) and glucose, fructose and particularly gluconic acid for *Aphanomyces cochlioides* (Rai and Strobel, 1966).

Chang-Ho and Hickman (1970) analysed root exudates from pea seedlings (3 weeks old), and detected the presence of low concentrations of a range of amino acids, organic acids, and glucose, fructose and sucrose. Glutamic acid, attractive at 0·75% (Royle and Hickman, 1964b), was present as 8·5 μg per plant. They measured the chemotactic attractiveness of anionic, neutral and cationic fractions of the exudate, of separated components of the exudate, and of pure chemicals at the same concentrations. They expressed these results in terms of the 'accumulation ratio'—the ratio of the number of zoospores adjacent to a capillary tube containing the extract and the number of zoospores randomly dispersed. The chemotaxis to the exudate could not be ascribed to the presence of any single chemical, and recombined fractions and complex mixtures were never as attractive as the crude extract (Table II).

Allen and Newhook (1973) have suggested that ethanol may be an important chemotactic attractant in root exudates. These authors, noting that infection of roots by *P. cinnamomi* is correlated with waterlogging of the soil, have reasoned that a concomitant increase in anaerobic fermentation by the roots would lead to a release of ethanol. They demonstrated that zoospores of *P. cinnamomi* showed no measurable change in swimming characteristics in solutions from 0 to 25 mM-ethanol, but were positively chemotactic to ethanol solutions (Table III). Three tests were used: the 'swim-in', in which the number of zoospores accumulating in a capillary tube containing the ethanol solution was counted; the 'swim-out', in which the number of zoospores remaining in distilled water in a capillary tube immersed in an ethanol solution was counted; and the 'rhizosphere model', which was devised to provide a permanent linear concentration gradient rather than the temporary

TABLE III. Chemotaxis of *Phytophthora cinnamomi* zoospores to ethanol[a]

Test	Number of zoospores in capillary tube	
	Water control	5 mM-ethanol
(a) 'Swim-in', ethanol in tube, after 30 min	22 ± 4	810 ± 330
(b) 'Swim-out', water in tube, 190 μm diameter, after 30 min	42	18
(c) 'Rhizosphere model', concentration gradient over 16 mm, tube 190 μm diameter, after 2 h	15	83
Percentage penetrated to 3 mm	52	90
to 6 mm	15	78
to 9 mm	0	52
to 12 mm	0	9

[a] Compiled from Allen and Newhook (1973).

gradient of the other two systems. In the rhizosphere model capillary tubes acted as diffusion bridges between a large stirred volume of the ethanol solution and, after equilibration, a large stirred volume of a zoospore suspension. After two hours, the numbers of zoospores penetrating different distances into the tubes were counted.

In the swim-in test, tactic responses were observed in the first 3–5 minutes, and the swimming zoospores showed a series of turning reactions and swam toward the source of ethanol. The tactic response became less with time, presumably as the concentration gradients flattened. In the swim-out tests, zoospores in the tubes swam towards the source of the ethanol at the ends of the tubes, and collided with the walls of the tubes less frequently, which might aid their passage between soil particles.

In the rhizosphere model the zoospores clearly responded to the concentration gradient of ethanol. Because of its volatility ethanol would not have been detected in earlier analyses of root exudates capable of attracting zoospores, but these results indicate that it could be an important attractant in anaerobic soils. Allen (1973) detected ethanol at a concentration of approximately 5 mM in radicles of *Lupinus angustifolius* grown in wet but well-drained soil, and in soil that had been saturated for 12–24 hours. Allen and Newhook (1974) showed that the spontaneous turning activity of the zoospores decreased systematically as the concentration of ethanol increased, thus resulting in more efficient chemotactic attraction.

Allen and Harvey (1974) have presented a detailed analysis of the negative chemotaxis of the *P. cinnamomi* zoospores to hydrochloric acid and its salts. They used the swim-out test, and also followed the rate of migration of bands of zoospores in capillary tubes away from the test solutions. The cells were found to respond to a critical absolute concentration of each chemical by acute, repetitive turning movements, and they swam away from the source of the chemical. With Cl^- as anion, these critical concentrations were calculated as 37 μM for H^+, 628 μM for K^+, 5040 μM for NH_4^+ and 6480 μM for Na^+. There was no evidence that an individual cell could respond to the concentration gradient over its surface, but instead the experiments suggest an all or none avoidance response at the boundary of the critical concentration. The relative threshold concentrations of the different cations followed the lyotropic series for cation exchange reactions, and Allen and Harvey suggest that a surface exchange of cations could differentially effect the surface charge of the two flagella. Zoospores encysted immediately when suspensions were mixed with acids or salts having a concentration higher than that which induced the turning reaction.

There are suggestions that the movements of some zoospores and possibly the growth of the germ tubes of leaf infecting fungi may be controlled by chemotaxis and chemotropism on the leaf surface itself (Flentje, 1959). Gregory (1912) and Arens (1929) described how zoospores of *Plasmopora viticola* aggregate and encyst around stomata on vine leaves and the germ tubes grow towards and through the pores to infect the leaf. Arens suggested that carbon dioxide could be involved in the attraction. Bald (1952) showed how important the formation of stomatal droplets containing exudates can be in the initial stages of infection, but it is not clear whether chemotropism is involved.

In classic accounts of fungi parasitic on other fungi Reinhardt (1892) and Burgeff (1924) described and illustrated hyphae of *Sclerotinia*, *Parasitella* and *Chaetocladium* being attracted by host hyphae from some distance away and either enveloping them or fusing with them. Germ tubes of *Piptocephalis virginiana* and *Dispira cornuta* show strong positive chemotropism to living host cells (Berry and Barnett, 1957; Barnett and Binder, 1973). Barnett and Binder (1973) also describe the inverse situation, with germinating spores of *Calcarisporium parasiticum* and *Gonatobotrys simplex* secreting attractants which specifically induce positive tropism of nearby host hyphae (*Sphaeropsis malorum* and *Alternaria* sp. respectively). They suggest that this unusual ability to cause tropism of the host is a survival mechanism, as these parasites do not produce long germ tubes. Related species that do produce long germ tubes do not show this ability.

Chemotaxis can also play a part in uniting symbiotic partners. *Platymonas convolutae* is the flagellated algal symbiont of the marine platyhelminth, *Convoluta roscoffensis*. Adult worms are always green but the symbiosis is not hereditary. Colourless eggs are laid and hatch to colourless larvae which have to encounter and engulf cells of *P. convolutae* in order to become green and develop. In the classic work on this symbiosis, Keeble and Gamble (1907) and Keeble (1910) described how motile cells of the symbiotic algae were attracted by *Convoluta* egg cases and settled on them; "we conclude, therefore, . . . that a definite substance diffusing out from the capsule-walls induces a tropistic (tactic) movement in the motile, algal cells of such a nature that they approach the source whence the chemical substance emanates." This chemotaxis has been confirmed using pure cultures of *P. convolutae*, the cells of which will cluster around both discarded and full egg capsules within five minutes (Holligan and Gooday, 1975). The nature of the attractant has not been investigated, but egg capsules remain attractive after being stored frozen overnight, and after washing with ethanol. Some of the cells settle on the egg capsules and divide. This alga, although a motile flagellate, shows a sedentary habit by remaining in pairs following cell division for a much longer time than its free-living relatives (Gooday, 1970) and so the cells accumulate on the egg capsules, and are well placed to be devoured by the hatching larvae.

IV. Social tropisms of fungi

A. *Observations of autotropism*

Fungal hyphae in a colony are patently aware of each other's presence. The most obvious manifestation of this is the circular form of a fungal colony on agar. The leading hyphae grow outwards approximately parallel to one another, and at approximately the same distance apart, so that new leading hyphae are added as the colony expands. Slower growing primary and then secondary branches are regularly formed further back in the colony, so that the mycelium fills the space available to allow efficient grazing through a primitive territorial system (Bonner,

1970). The density of growth of individual hyphae to form the colony is controlled by the availability of nutrients. The outer hyphal tips are further apart on a poor medium, or when the colony crosses a bridge of cellophane or water agar, but they are still approximately equidistant. This cannot be explained by each hypha affecting the others' direction of growth by a localized depletion of nutrients, for as discussed above, fungal hyphae show little or no directional response to nutrients. We must assume that the hyphae have other means of external and internal co-ordination to control their growth.

Evidence for external chemical control of neighbouring cells comes from studies of orientation of germ tube emergence and growth with fungal spores. The work of Clark (1902), Fulton (1906) and Graves (1916) established the negative autotropism of fungal germ-tubes. These and later authors used the technique of Miyoshi (1894) in which spores are spread on a medium separated from another medium by a perforated plate. When the other medium is uninoculated the germ-tubes markedly grow towards the perforations. When the other medium is equally inoculated, a random orientation is seen. Fulton (1906) observed less orientation towards staled medium than towards fresh medium, and Graves (1916) observed preferential orientation away from staled medium which had been inoculated with spores, so the presence of a negatively autotropic "staling substance" was suggested. This was investigated in detail by Stadler (1952, 1953), chiefly with *Rhizopus nigricans*. Stadler made measurements of angles of growth of germ-tubes on agar medium with respect to a 1 mm diameter hole in a sheet of plastic separating them from the opposite layer of agar, an angle of 0° indicating growth directly towards the hole and an angle of 180° growth directly away from the hole. When germinating spores were on both sides, a random orientation with a mean at 92° was recorded. With spores on one side and nutrient or water agar on the other side, oriented growth towards the hole was recorded as mean angles of 46° and 44° respectively. Thus the germ-tubes are growing away from each other rather than towards a source of nutrient. There was no significant difference in the growth towards nutrient, staled growth medium, boiled staled growth medium or water agars, so Stadler concluded that the autotropic factor was labile and did not accumulate in staled medium. As well as opposed growth of germ-tubes, another manifestation of the negative autotropism was that the sites of germ-tube emergence tended to be at opposite ends of closely placed spores. Thus the germ-tubes from a clump of spores would be produced radially outwards. Spores germinating in flowing medium preferentially grew upstream, and Stadler interpreted this as reflecting the higher concentration of the autotropic metabolite downstream. Stadler discussed two hypotheses as to the possible mechanism of action of the autotropic factor.

1. *The growth factor hypothesis*

He suggested that each cell produces a growth stimulant which diffuses into the medium to give a higher concentration on the near side of an adjacent germ-tube which then responds by bending away through enhanced intercalary growth.

2. *The wall strengthener hypothesis*

He suggested that each cell produces a factor which causes wall strengthening, and that the limiting factor in the growth of a germ-tube apex or the bulging of a swelling spore at the onset of germ-tube emergence is the plasticity of the germ-tube or spore wall. Thus swelling on the far side of an adjacent germ-tube or spore will be inhibited less than on the near side and negative autotropism with respect to both germ-tube emergence and growth result. Since Stadler observed that germ-tube growth was strictly apical and not intercalary, he favoured the wall strengthener hypothesis which is consistent with apical extension.

As the autotropic factor could be demonstrated in the experiment with flowing medium but not by transferring staled medium, and as its action was apparently not affected by different aeration conditions, Stadler concluded that it was an unstable non-volatile metabolite.

Jaffe (1965, 1966) and Müller and Jaffe (1965) rigorously investigated autotropism, by comparing the behaviour of germinating spores of *Botytis cinerea* with theoretical models predicting the diffusion of factors in different conditions. One set of experiments involved the orientation of germ tubes of spores electrostatically precipitated sparsely onto a polythene film and then subjected to medium flowing past at different. rates. Contrary to Stadler's result with higher concentrations of spores of *R. nigricans*, these germ-tubes preferentially grew downstream, and this tendency increased with increasing flow rates. In multi-chambered experiments, the presence of high concentrations of spores upstream reduced the preferential downstream orientation. After considering several possibilities, such as effects of molecular orientation, wetting and pressure, Müller and Jaffe concluded that their observations could be explained by the production by the cells of a macromolecular diffusible growth stimulator, with a half-life of about 10 seconds, that would be washed downstream from the spores.

Jaffe (1966) then investigated the autotropism of adjacent pairs of germinating spores of *B. cinerea*. He statistically analysed data of the configurations shown by the two emerging germ-tubes by classifying them as *cis* (on same side of spore) or *trans* and as (++) (towards one another), (+−) or (− −). There was a strong tendency for the *cis* (++) configuration of positive autotropism, with the germ-tube emerging towards the neighbouring spore and then growing towards it. If the carbon dioxide concentration was raised to 0·3% or 3% the spore pairs showed negative autotropism, with a predominent *cis* (— —) configuration. Using the same recording procedure, Robinson *et al.* (1968) investigated the behaviour of spore pairs of *Rhizopus stolonifer*, *Mucor plumbeus*, *Trichoderma viride* and *B. cinerea* on agar media and on cellophane over the agar. Spores touching each other or separated by less than 40 μm showed a tendency for *cis* configurations. The first three species, particularly the two Mucorales, also showed a tendency to negative autotropism, but *B. cinerea* appeared autotropically neutral. However, these authors reported that they confirmed Jaffe's observation of positive autotropism of electrostatically precipitated spore pairs, and emphasised that autotropic behaviour varies under different experimental conditions. They also reported that touching spore pairs of *R. stolonifer* and *M. plumbeus* germinated more quickly than single spores, a result consistent with the observation by Müller and Jaffe (1965) that spores of *B. cinerea* germinated more slowly in flowing medium.

Robinson (1973a, b, c) investigated autotropism of germinating spores of *Geotrichum candidum*, and recorded patterns of germ-tube emergence in spore pairs and tropism in perforated plate experiments and in spores germinating under coverslips. Groups of two or three spores germinated more quickly than single spores, and the negatively autotropic germination pattern of spore pairs was very marked, with germ-tubes always emerging from opposite ends of the two spores. In perforated plate experiments with uninoculated media on the opposite side of the hole the response varied with the density of the inoculum. At high concentrations (above 10^6 spores/ml) the spores germinated towards the hole, but at low concentrations (10^5 spores/ml or less) there was no significant effect on the site of germ-tube emergence but the resultant germ-tubes ultimately grew towards the hole. Spores inoculated under a coverslip germinated predominantly towards its edge at the higher concentrations, but not at the lower concentrations. Robinson's experiments demonstrate three effects; a germination promotion between neighbouring spores, a negative autotropism of site of germ-tube emergence, and a negative autotropism of growth of the germ-tube. No evidence was obtained for positive autotropism. The first two effects were seen only over a very short range, while the third was observed over a long range. Robinson suggested that the autotropic responses could be positive chemotropism towards oxygen, by growth up a concentration gradient, and re-interpreted Stadler's observations of negative autotropism of *R. nigricans* as positive chemotropism to oxygen. Germinating spores in a perforated plate experiment would lower the oxygen tension on their side of the plate and so give rise to an oxygen gradient towards the hole. Robinson eliminated the possibility that the autotropism represents negative chemotropism to carbon dioxide by experiments showing no differences in orientations in perforated plate experiments in different atmospheres of carbon dioxide which would be expected to give different gradients across the plates. Direct tests on the chemotropism of fungal hyphae to oxygen are needed and should be possible if artificial oxygen gradients can be constructed in volumes sufficient to minimize any changes in concentration brought about by the cells. Chemotaxis to oxygen is widespread in bacteria (Barrachini and Sherris, 1959; see also Ch. 3, this volume) and Myxomycete plasmodia under partially anaerobic conditions will migrate towards an oxygen source (M. J. Carlile, personal communication).

An important property of many Ascomycete and Basidiomycete fungi is their habit of forming vegetative hyphal fusions following specific inter-hyphal attractions. The fusions can be between cells of the same or different mycelium or mating type, and can give rise to sterile or fertile heterokaryons, a dikaryon, or involve no nuclear exchange. Such fusions are rare or absent in the Phycomycetes, except for a few genera such as *Endogone*, and here are considered separately from the specific sexual fusions discussed later. The phenomenon has intrigued many mycologists, notably Buller (1931, 1933) who made extensive observations on fusions in many different fungi. The formation of hyphal fusions means that mycelium from either one spore or several neighbouring spores becomes converted into a three-dimensional network, and "competition becomes co-operation". Buller lists several functions of hyphal fusions: to convert young mycelium into a network to give conduction of food materials in any direction; to allow the mating of mycelia and to allow passage of nuclei through a haploid mycelium becoming dikaryotized (see below); to enable mycelium to resist mechanical injury so that if one hypha is cut the mycelium is not bisected, and to repair wounds by bridging gaps; to allow social organization so that many mycelia of the same species can unite to form a single compound mycelium to give a fruit body; and to allow the formation of nematode traps made of loops of hyphae. In addition, hyphal fusions within fruit bodies allow for easier translocation and co-ordination and increase the mechanical stability. Hyphal fusions are present in fruit bodies of Ascomycetes and Hymenomycetes as described by Buller, and also in much simpler structures such as the coremia of *Penicillium* species (Carlile *et al.*, 1962), and the microsclerotia of *Pleiochaeta setosa* (Harvey, 1975; Fig. 1). The formation of these microsclerotia involves the anastomosis of hyphal buds, and of these buds and other hyphae. There appears to be a clear directional response of the buds towards the point of fusion (Fig. 1).

There seems to have been a reluctance to accept the involvement of chemotropism in hyphal fusions despite the fact that action at a distance during hyphal fusion was beautifully described by Ward (1888) in

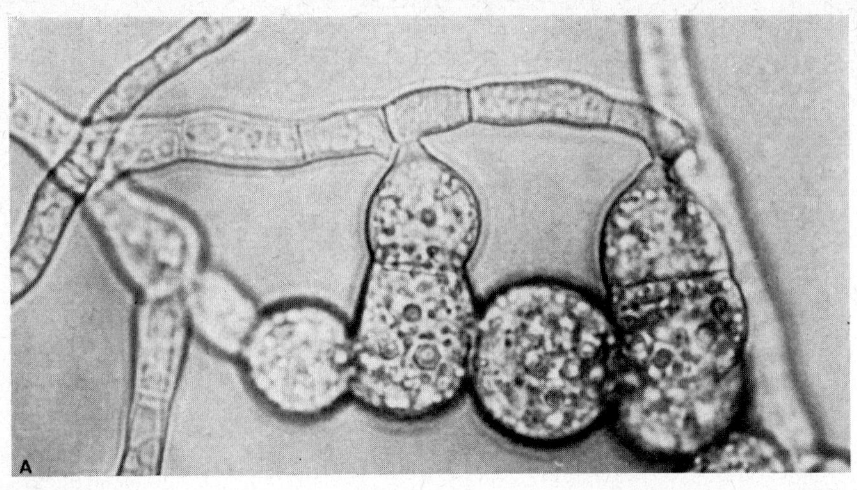

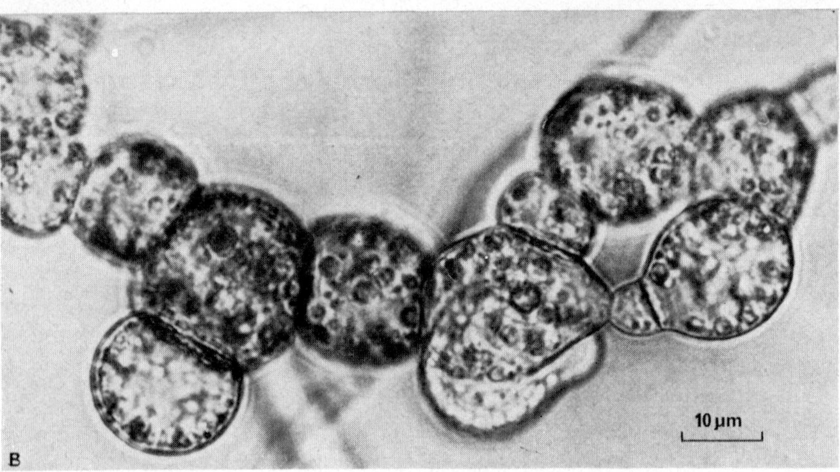

FIG. 1. Cell fusions during formation of microsclerotia of *Pleiochaeta setosa*. The fungus is growing embedded in agar. (A) shows fusions of buds to a neighbouring hypha; and (B) shows a fusion between adjacent buds, which have elongated towards one another. Magnification ×1100. Photographs by courtesy of Dr. I. C. Harvey.

Botrytis. He concluded that "it seems to me impossible to avoid the impression that some attraction is exerted". Buller classifies four kinds of fusion: hypha-to-hypha fusions, between two hyphal tips that attract each other from about 15 μm apart; hypha-to-peg fusions, between a hyphal tip and a peg produced as a side branch from an older hypha in response to this advancing hyphal tip (Fig. 2); peg-to-peg fusions, between two pegs produced opposite one another from adjacent hyphae (Fig. 3); and the hook-to-peg fusions during clamp connection formation, discussed later. All these hyphal fusions involve the chemotropism

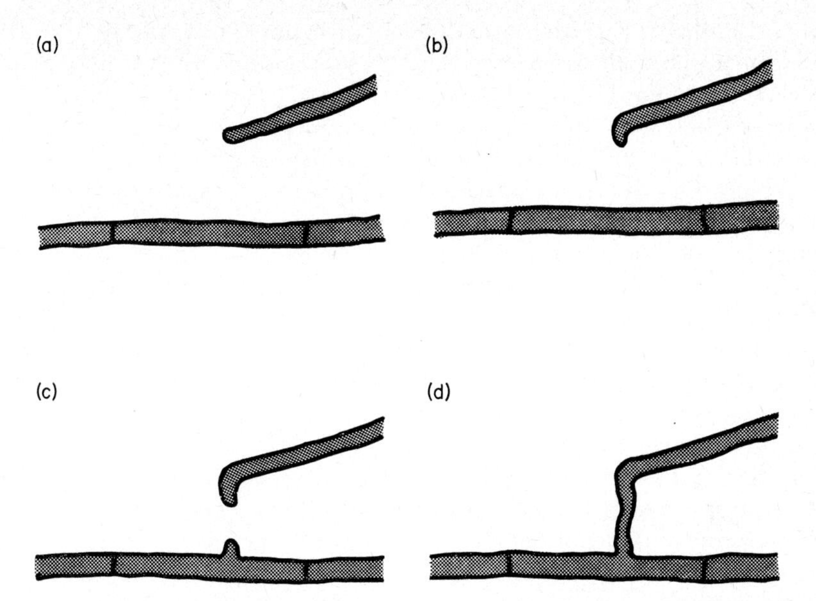

Fig. 2. Generalized diagram showing successive stages (a–d) in a hypha-to-peg fusion in the higher fungi. After Buller (1933).

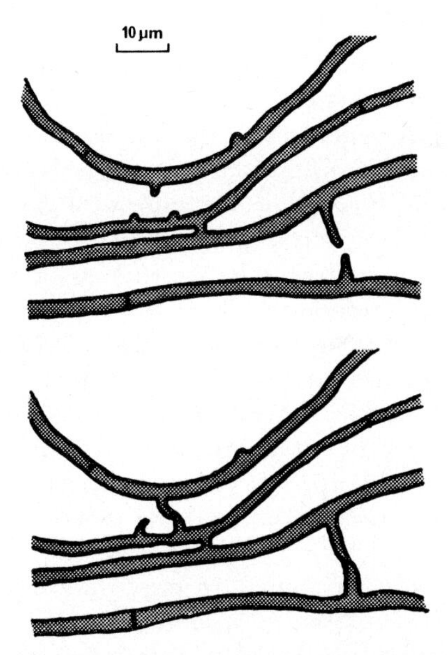

Fig. 3. The formation of two peg-to-peg fusions in *Pleurage curvicolla*. Each of the two lower hyphae produced a peg, and these grew to each other and fused a few minutes later. However, in the upper pair of hyphae, one had produced two pegs, both of which grew towards the peg produced by the topmost hypha. Only one of these two pegs fused, whereupon the unsuccessful peg stopped growing. The scale line is 10 μm. After Buller (1933).

of both fusing cells, either as a directed growth of a hyphal tip or in the case of peg fusions as the localized formation *de novo* of a hyphal tip which then shows directed growth. Ward (1888) and Buller (1933) observed that sometimes two pegs were formed by the one hypha under the influence of an approaching hypha or peg, and that if only one peg fused successfully the unsuccessful peg quickly stopped growing, as if the growth stimulus ceased to be formed after the two tips had fused (Fig. 3). A similar cessation of growth has been observed during the attraction of sexual hyphae in *Mucor mucedo* (Fig. 7, discussed later).

Köhler (1929, 1930) reported that hyphal fusions are very specific, as they only occurred between cells of the same species, although the chemotropism was observed between closely related species. Unrelated species showed no reaction.

Buller (1933) discusses the possible mechanisms for the mutual chemotropism of two vegetative cells of the same species. He dismisses the possibility that only one chemical is involved as then there would be no gradient from one cell to the other. He also admits to the danger of suggesting two complementary substances, one from each hypha, as this might lead to a *reductio ad absurdum*, with every hypha characterized by a different chemical, but he does end by suggesting that two chemicals are involved, the production of only one in each hypha being controlled by an unknown mechanism. Raper (1952) restated this problem of what basic and common type of mechanism could control these species-specific hyphal fusions, regardless of the origin of the two cells and of the nature of the product of the fusion. He suggested that every growing cell produces a single species-specific substance that is labile, and has a steep concentration gradient over about 10 μm. Hyphal tips and hyphae of the same species would be sensitive to a critical concentration of this substance, and respond by growing towards its increased concentration and by producing a new growing tip which would behave in the same way. This idea is expounded and illustrated by Burnett (1968). Park and Robinson (1966) suggest that hypha-to-hypha fusions could be controlled by the negative chemotropism of the tips away from a staling factor, by assuming that tips can remove this factor. Two tips would then converge. This suggestion does not explain the formation and fusion of pegs.

The control of hyphal fusions is an important aspect of fungal morphogenesis in which it is essential to obtain quantitative measurements before further advances can be anticipated. Ahmad and Miles (1970) have shown the importance of counting hyphal fusion frequencies in their demonstration of the involvement of mating-type genes in hyphal fusions in *Schizophyllum commune* (see below).

B. *Possible mechanisms of autotropism*

How many different mechanisms are involved in the autotropisms of fungi? There are several different effects observable, fungal spores germinating and growing towards or away from each other and fungal hyphae growing away from each other to form a circular colony or towards one another to give hyphal fusions. Most of the observations can be explained by assuming two basic mechanisms; a long range non-specific effect and a short range species-specific effect.

The long range non-specific effect would be the tropism either away from "staling factors" as suggested by Stadler (1952, 1953) or towards oxygen as suggested by Robinson (1973a, b, c). This tropism would be responsible for the radial outgrowth and spacing of hyphae in a colony and for the outward germination and growth of spores. Gradients of oxygen or staling factors could develop over very short or very long distances and could vary greatly with different conditions.

The short range species-specific effect would be a growth stimulation and positive chemotropism responsible for the growing together of *Botrytis* spores described by Jaffe (1966) and Müller and Jaffe (1965), and for the hyphal fusions seen in Ascomycetes and Basidiomycetes. Vegetative hyphal fusions are not seen in the Mucorales, and so these fungi might lack this mechanism. This would explain why Stadler (1952, 1953) and Robinson *et al.* (1968) using *Rhizopus* and *Mucor* spores did not see the effects observed by Jaffe (1966). Jaffe used *Botrytis*,

TABLE IV. Comparison of properties suggested for the attractant of *Botrytis*, and the attractants controlling hyphal fusions in higher fungi[a]

	Botrytis attractant	Hyphal fusion attractant
Specificity	?	Species specific
Molecular size	Macromolecule, 10^5–10^8 Daltons	Large enough for specificity
Stability	Unstable, half-life 10 sec	Unstable
Radius of action	*Ca.* 7 μm	*Ca.* 10 μm
Site of production and action	Germ-tube apex, point of germination	Hyphal apex, point of peg development
Chemotropic response	Positive	Positive

[a] Compiled from Jaffe (1966), Müller and Jaffe (1965), Raper (1952), and Buller (1933).

which is well known for its prolific hyphal fusions since the classic work of Ward (1888). Table IV compares the properties deduced by Müller and Jaffe (1965) for the growth stimulant and attractant of *Botrytis* and by Raper (1952) for the attractant controlling fusions. Müller and Jaffe's calculations indicate a macromolecule with a half-life of about 10 seconds. From immunological and other work it is clear that there are specific proteins and glycoproteins on the surface of and diffusing out from fungal cells. If one of these were the putative attractant its instability could be caused by several possible mechanisms—it could be released together with a degrading enzyme, its conformation could be inherently unstable or could be altered by the loss or gain of an effector molecule in the medium, or it could be an oligomeric macromolecule that dissociated into two or more sub-units at a lower concentration. This last suggestion is attractive in that such sub-units could re-aggregate with sub-units from a neighbouring cell to enhance the concentration of active polymer between the two cells.

V. Sexual taxes and tropisms

Sexual chemotaxes and chemotropism occur in many fungi and algae, but this account will be restricted to those where some details of the chemical or physiological mechanisms are known and may be of wider significance. Other examples are reviewed elsewhere (Köhler, 1967; Machlis and Rawitscher-Kunkel, 1963, 1967; Raper, 1967; Wiese, 1969; Ziegler, 1962a, b).

A. Algal sex attractants

1. *Chlamydomonas*

Sexual chemotaxis between gametes of the green alga *Chlamydomonas* was investigated in the 1930's by F. Moewus as part of an extensive study of sexuality; much of this work was not repeatable by other workers—see "*in memoriam*" in Raper (1957). There is, however, one convincing report of unidirectional sexual chemotaxis in *Chlamydomonas moewusii* var. *rotunda* by Tsubo (1957, 1961). Tsubo showed that (+) gametes of this strain are attracted by (—) gametes or by capillary tubes containing medium from (—) cultures. Chemotaxis was not observed with the same technique in four other species of *Chlamydomonas* and has not been reported for any other species by other workers. These other species presumably rely on agglutinations between gametes that have aggregated by another stimulus, such as phototaxis. However, the attractant was not species specific as both mating types of *C. moewusii*, *C. eugametos* and *C. moewusii* var. *tenuichloris*, but not *C. reinhardi*, showed positive chemotropism to the culture filtrate of (—) *C. moewusii* var. *rotunda*. Tsubo showed that the attractant is volatile as it was quickly lost from the culture filtrate by bubbling with nitrogen and so it is possible that it could be characterized by the techniques used by Müller and Jaenicke for ectocarpen and fucoserraten. This attractant from *C. moewsii* var. *rotunda* was shown to be a specific sex attractant for this species as it was only detected from sexually active (—) cultures and only active on sexually active (+) cultures. Coal gas, ethylene and ethane could mimic the effect and specificity of the sex attractant, but the relative concentrations required are not clear.

Although Wiese (1969) stresses the clear distinction between chemotaxis as described by Tsubo, and the sexual agglutination of *Chlamydomonas* mediated by high molecular weight glycoproteins, it is possible that the agglutinins could have a short range attraction (Brown *et al.*, 1968). As described by Wiese (1965, 1969), the glycoproteins extracted from both mating types have the property of isoagglutinating gametes of opposite mating type so that they clump together, just as *Ectocarpus* male gametes do when treated with ectocarpen (see below). Brown and co-workers observed streamers of material that they identify with the agglutinins diffusing out from the flagellar tips of gametes and suggest that these might exert an attraction just before contact of two gametes.

2. *Ectocarpus and ectocarpen*

Sexual reproduction in the small brown seaweed *Ectocarpus* involves the release of isomorphic biflagellate male and female gametes. The female gametes swim and attach themselves to the substrate by their anterior flagella. Each then strongly chemotactically attracts the male gametes so that they cluster around. After fertilization the zygote no longer attracts the male gametes. Thus the chemotactic agent is only released by the female gametes after they have settled and before they are fertilized. Müller (1967a, b; 1968) has shown that it is possible to obtain an active extract of a volatile agent in a cold-trap after bubbling air through a female culture of *Ectocarpus siliculosus* that is producing gametes. The resulting material has two powerful effects on suspensions of male gametes; clumping and attraction. When a drop containing gametes is exposed to the vapour their swimming behaviour alters so that they aggregate in small clumps. When a small column of a gamete suspension in a glass tube is exposed on one side to the vapour and on the other to air the gametes congregate at the meniscus exposed to the active material (Müller, 1967b, 1968, 1972a). The active material can also be trapped in a drop of mineral oil, which will then attract the male gametes (Fig. 4).

By trapping the active material over two years 92 mg was obtained from 1041 g of fresh gametophyte material. Chemical analysis, using gas chromatography, mass spectrometry and spectroscopy, has shown that the active molecule "ectocarpen" (Fig. 10) is the hydrocarbon $C_{11}H_{16}$, with three non-conjugated *cis* double bonds and a seven carbon ring (Müller *et al.*, 1971; Jaenicke, 1972a, b). This structure has been confirmed by synthesis (Jaenicke *et al.*, 1971). Nothing is known of its biosynthesis, but Jaenicke (1972b) and Jaenicke and Müller (1973) suggest two possible schemes, either from linolenic acid by β-oxidation and then α-oxidation to give the C_{11} aldehyde which could then cyclise, or from a C_{10} polyene, *cis*-decatetraene, with addition of a methyl group

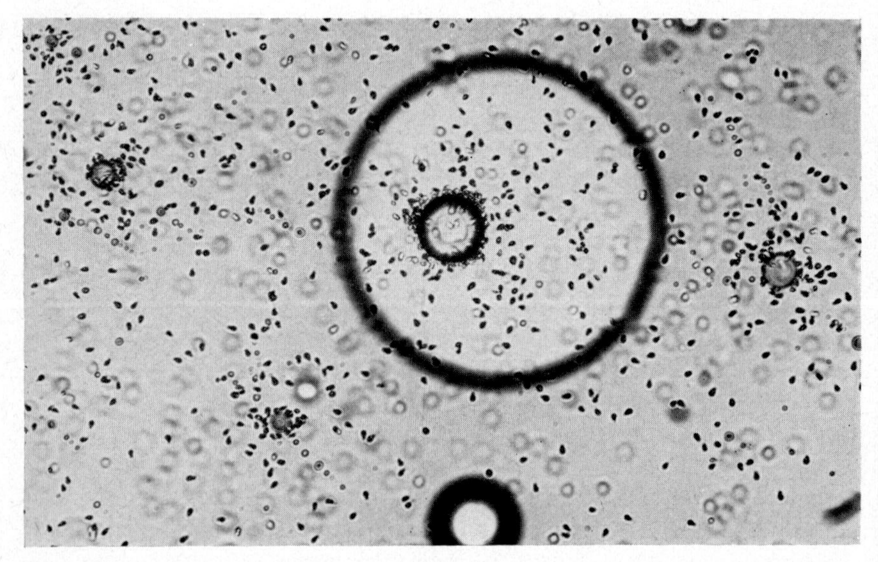

FIG. 4. Chemotaxis of male gametes of *Ectocarpus siliculosus* to droplets of paraffin oil containing ectocarpen that have been suspended in the medium. The diameter of the small central drop is 50 μm. Photomicrograph provided by Dr. D. G. Müller.

to give a cyclopropane ring and then a rearrangement. The latter scheme gains some credence as a similar C_{11} hydrocarbon with a cyclopropane ring has been characterized from another brown alga, *Dictyopteris plagiogramma* (Pettus and Moore, 1971). No trace of a cyclopropane ring has been detected in material from *E. siliculosus* (Jaenicke *et al.*, 1971), but an intermediate need not accumulate in detectable amounts. Jaenicke *et al.* (1974) suggest that *cis*, *cis*-undeca-1, 5, 8-trien-3-ol could be the common precursor for ectocarpen, multifidene, and fucoserraten.

Ectocarpen appears to be a specific attractant for *Ectocarpus* male gametes. The male gametes of a closely related species, *Giffordia mitchellae* are not attracted by it (Müller, 1972a) although they are attracted by their own settled female gametes (Müller, 1969).

3. *Fucus and fucoserraten*

There have been many reports of the attraction of male gametes to eggs and oogonia of algae, but only recently has progress been made in identifying the compounds involved. It has been particularly satisfying that the classic work of Cook *et al.* (1948), and of Cook and Elvidge (1951) has now come to fruition. These authors made considerable progress by obtaining volatile fractions with biological activity by aspirating suspensions of the eggs of the brown seaweeds *Fucus serratus* and *Fucus vesiculosus* with a stream of nitrogen or hydrogen. These fractions were trapped in liquid air, and strongly attracted the sperms which responded to the concentration gradient of the active material. Bioassays were carried out by placing capillary tubes containing the extracts into a suspension of sperms in seawater. The swimming cells aggregated around the mouths of the tubes containing the active extracts. The active material was analysed by the most sophisticated techniques then available, particularly by mass spectrometry. Cook and Elvidge suggested that it was a hydrocarbon and showed that many hydrocarbons and other simple organic molecules could attract the sperms in the bioassay, but only at much higher concentrations than the natural product.

Müller and Jaenicke have now re-investigated this system. Müller (1972b) has isolated the material in a cold trap and purified it by gas chromatography to obtain 690 μg from 252 kg of fresh female receptacles. He has devised a simple bioassay by exposing a small piece of glass with spots of vaseline grease to an air-stream to be tested. The piece of glass is then placed alongside a control piece in a sperm suspension. The sperms rapidly gather around the vaseline if it has absorbed the active compound (Fig. 5). The purified material was analysed by mass spectrometry, spectroscopy and gas chromatography of the parent compound and of its hydrogenated derivative and shown to be a conjugated triene hydrocarbon, 1, 3, 5-octatriene, 'fucoserraten' (Müller and Jaenicke, 1973). The stereochemistry has been established as *trans*, *cis*-1, 3, 5-octatriene by synthetic work (L. Jaenicke and K. Seferiadis, quoted in Jaenicke *et al.*, 1974).

4. *Cutleria and multifidene*

A third hydrocarbon, multifidene, is responsible for the attraction of the male microgametes to the female macrogametes in the brown seaweed *Cutleria multifida* (Müller, 1974; Müller and Müller, 1974; Jaenicke *et al.*, 1974). Multifidene, $C_{11}H_{16}$, contains a five carbon ring (Fig. 10). The female macrogametes also produce two biologically inactive minor co-metabolites with the same molecular formula, one with a six carbon ring and the other the seven carbon ring, ectocarpen.

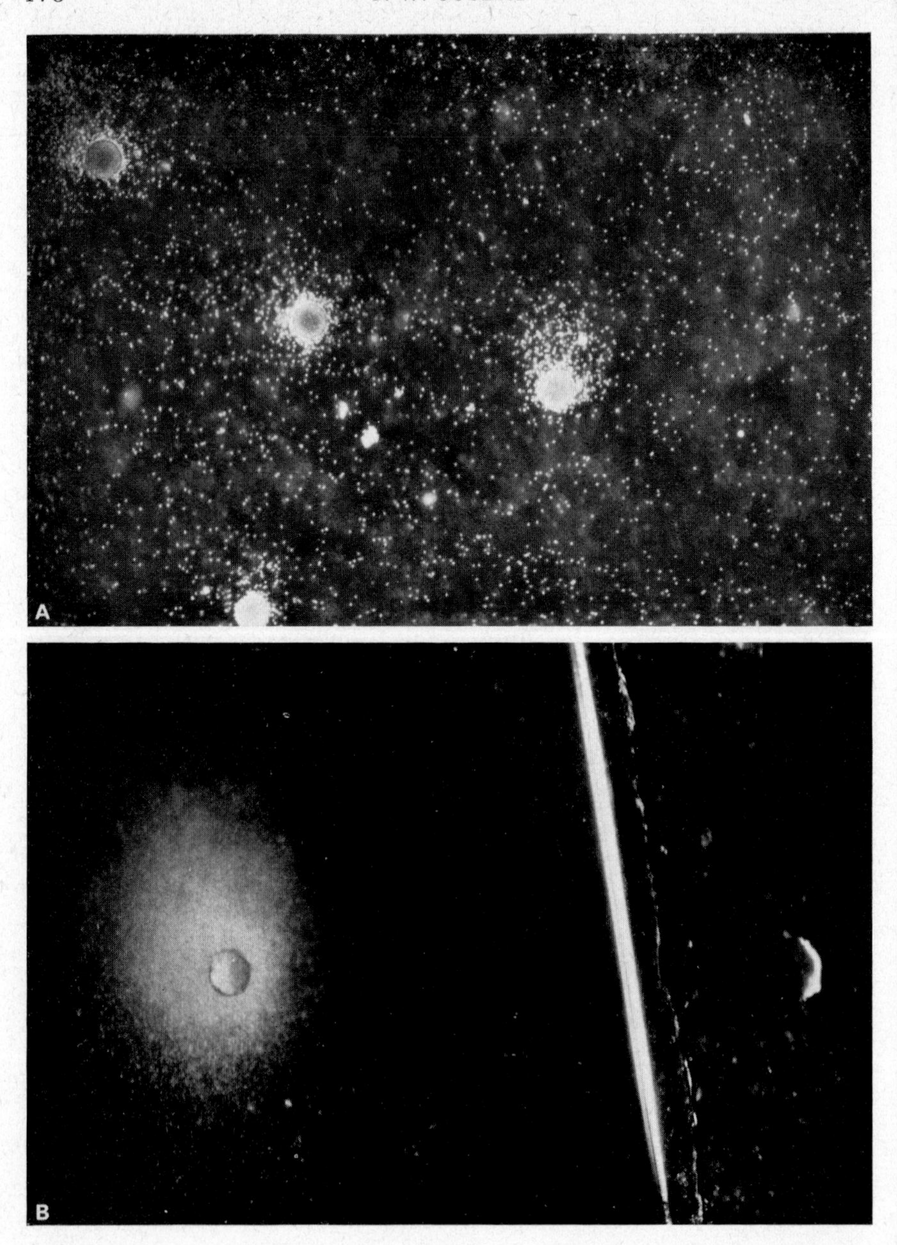

FIG. 5. Chemotaxis of spermatozoa of *Fucus serratus*. (A) Spermatozoa clustering around live eggs, phase contrast microscopy. The egg diameter is 70 μm. (B) Spermatozoa clustering around a drop of Vaseline on the left hand side that has absorbed purified fucoserraten from the outlet of a gas chromatograph. No spermatozoa are around the control Vaseline drop on the right hand side. Dark field microscopy. The Vaseline drops are about 200 μm in diameter. Both photographs provided by Dr. D. G. Müller.

Jaenicke *et al.* (1974) suggest a common biogenesis for these metabolites from the C_{11} polyene alcohol, *cis*, *cis*-undeca-1, 5, 8-triene-3-ol. They further suggest that fucoserraten could also be derived from this compound by oxidation at C-7 and heterolytic fragmentation.

5. *Oedogonium*

Progress has also been made with the study of the chemotaxis of male cells towards female cells of *Oedogonium*, a filamentous green alga. Chemotaxis in this alga was suggested by observations made more than 100 years ago (see Machlis *et al.*, 1974), as sperms can be seen to cluster around oogonia. Hoffman (1960, 1961) investigated fertilization in *Oedogonium cardiacum* and another heterothallic species in both of which the sperms are produced in antheridia on the male plant. He tested aqueous extracts of homogenized female plants, and the supernatant from plants producing oogonia. Both extracts attracted sperm cells when added to sperm suspensions in a capillary tube. The biological activity remained in the aqueous phase after extraction with ether and chloroform. The activity appeared species-specific, as there was no appreciable cross-reaction between the two species used, and the extracts were only active when made from fertile filaments. This suggests that the biosynthesis of the active material is controlled by the differentiation that gives oogonia.

Some other species of *Oedogonium* are 'nannandrous', the male filaments producing androspores instead of directly producing sperms. The androspores swim to female filaments where they attach themselves, and differentiate to produce the sperms which are released close to the oogonium. Hoffman (1961) has shown that androspores of *Oedogonium idioandrosporum* are attracted by extracts of filaments bearing oogonial mother cells, but not by extracts from vegetative or oogonial filaments, or by sperm-attracting extracts of *O. cardiacum.*

Hormonal control of sexual reproduction has been investigated in another nannandrous species, *Oedogonium borisianum*, by Rawitscher-Kunkel and Machlis (1962). The androspores are chemotactically attracted by the oogonial mother cells by an agent that is produced by the female cells until androspores attach to them. The subsequent directional growth of the dwarf male and the division of the oogonial mother cell must be chemically controlled by the female and male plant respectively. Finally the sperms are attracted to the oogonia as in *O. cardiacum.*

Machlis *et al.* (1974) have investigated the chemical nature of the sperm attractant of *O. cardiacum.* Axenic cultures of the alga failed to reproduce sexually unless bacteria, especially *Pseudomonas putida*, were added to them. Sperm suspensions and oogonial cultures were then produced by transferring washed plants to nitrate-free medium. The sperm suspensions were used for the bioassay. Machlis had earlier developed a bioassay for sirenin by counting cells that settled on a membrane separating them from the test solution, but this did not work for *Oedogonium*. Therefore, Machlis and his colleagues designed a bioassay in which the behaviour of sperm cells is observed when they are added to wells between agar plugs, that contain the extracts to be tested and control water-agar plugs. The cells cluster next to active extracts within 2 to 3 minutes, and the concentration of the active material can be estimated by determining how many times it can be diluted before eliciting no response. The active culture filtrate from oogonial plants was concentrated by ultrafiltration with a filter passing molecules of molecular weight below 500. Gel filtration with Sephadex G-25 and G-15 indicates a molecular weight between 500 and 1,500. Activity during purification stages, including gel filtration, ion exchange and paper chromatography, has so far been associated with a yellow pigment with a characteristic ultra-violet spectrum. Machlis and his colleagues conclude that the *O. cardiacum* sperm attractant is a relatively high molecular weight yellow pigment, soluble in water but not in organic solvents.

B. *Fungal sex attractants*

1. *Allomyces and sirenin*

Sirenin is the female sex attractant of the water mould *Allomyces*. Its existence was demonstrated by Machlis in 1958, and its structure was reported ten years later. Haploid plants of *Allomyces* produce male and female gametangia, usually in pairs. When mature, the motile male and female gametes are released. The smaller male cells, bright orange with γ-carotene, swim actively and fuse with the more sluggish colourless females to form diploid motile zygotes. Machlis (1958a, b, c) showed that the female gametes synthesise a metabolite 'sirenin' that attracts the male gametes by chemotaxis. Sirenin can be assayed by counting the number of male gametes attracted to the surface of a cellophane membrane separating them from the test solution. The number of sperm settled on the membrane increases for about an hour and a half and then declines. The optimum conditions for such parameters as sperm concentration and time have been investigated by Machlis (1968, 1973a) and now the assay can detect a 50 picomolar (pM) solution of sirenin.

The production of sufficient sirenin for characterization was accomplished by Machlis *et al.* (1966). The growth of large volumes of the fungus had to be designed to obtain the maximum yield by synchronizing the release of large numbers of female gametes. The presence of significant numbers of male gametes had to be avoided, as they would inactivate the sirenin and also give rise to zygotes and a culture overrun with the resulting diploid plants. Thus the fungus used was not a wild-type *Allomyces*, but a predominantly female hybrid between two species. This hybrid had earlier been produced by Emerson and Wilson (1954) during study of speciation in these fungi. A predominantly male hybrid was used to provide the sperm suspensions for the assay.

Sirenin (Fig. 10) proved to be a bicyclic sesquiterpenediol (Machlis *et al.*, 1968; Nutting *et al.*, 1968). Its structure and stereochemistry have been confirmed by synthesis (Plattner *et al.*, 1969; Bhalerao *et al.*, 1970; Plattner and Rapoport, 1971). The bio-synthesis of sirenin almost certainly proceeds by cyclization of farnesyl pyrophosphate, but has not been investigated in detail.

Machlis (1973b) examined isomers and analogues of sirenin for biological activity. Only *l*-sirenin is active and other isomers do not show antagonism when added with *l*-sirenin. Machlis (1973a) also showed that chelated trace metals and 3 mM calcium are required for a maximal response to sirenin. The male gametes specifically inactivate sirenin and Carlile and Machlis (1965a) suggested that this could be a mechanism whereby they maintain their sensitivity over a very wide concentration gradient (10^{-10} to 10^{-5} molar). Machlis (1973a) showed that this up-take of sirenin follows first order kinetics with 5–400 nanomolar (nM) solutions. The sirenin cannot be extracted back from the spores, and so must be metabolized to an inactive product. The gametes lose their ability to respond to sirenin after taking it up; a period of 45 min was required to regain the full response from gametes that previously been 30 min in 5 nM-sirenin.

Five different motile cells are formed during the life cycle of *Allomyces*—haploid and diploid zoospores, male and female gametes, and zygotes. The chemotactic responses of these cells are shown in Table V. Only

TABLE V. Chemotactic responses of motile cells of *Allomyces*[a]

Motile cell	Production of sirenin	Response to sirenin	Response to amino acids
Male gamete	0	+	0
Female gamete	+	0	0
Zygote	0	0	+
Haploid zoospore	0	0	+
Diploid zoospore	0	0	+

[a] From Carlile and Machlis (1965b); Machlis (1969a, b).

the male gametes respond to sirenin, but they lose this response after fertilization and the resulting zygote very quickly gains the response to amino acids (Machlis, 1969b). Male gametes of different species of

Allomyces show considerable differences in sensitivity to sirenin (Machlis, 1968), but the most sensitive respond to a solution of 22 pg/ml compared with the 400 μg/ml casein hydrolysate required for the chemotaxis of zygotes (Carlile and Machlis, 1965b). The zygotes and haploid and diploid zoospores germinate by settling and producing germ-tubes, and Carlile and Machlis (1965b) showed that these germ-tubes exhibit clear positive chemotropism to casein hydrolysate.

2. *Achlya and antheridiol*

The control by hormones of sexual reproduction in the water mould *Achlya* has been studied by Raper (1939, 1940) and Barksdale and colleagues (see reviews by Raper 1970, 1971; Barksdale, 1967, 1969; Gooday, 1972, 1974). Raper showed that female strains produce diffusible factors that cause male strains to undergo sexual differentiation and form antheridial branches. These sexual hyphal branches are then chemotropically attracted by diffusible factors from the female. We now know that female strains synthesise and release a steroid hormone, antheridiol (Fig. 9), which added to unmated male strains causes the production of antheridial branches, the first visible stage in sexual reproduction. The purification and characterization of antheridiol from female culture filtrates of *Achlya* was a triumph of collaboration between biologists and chemists (McMorris and Barksdale, 1967; Arsenault *et al.*, 1968). Using purified antheridiol, Barksdale (1963a, 1967, 1969) has been able to show that at increasing concentrations antheridiol can elicit the complete developmental sequence of the male sexual hyphae. By absorbing antheridiol onto plastic particles near male hyphae she has shown that the resulting antheridial branches grow towards the source of antheridiol, wrap themselves around the particles and delimit the antheridia, in which meiosis then occurs. These "plastic oogonia" apparently mimic the female oogonia that would be the source of antheridiol at this stage of natural sexual reproduction. Another effect of antheridiol is the induction of the synthesis by male cells of 'hormone B', which diffuses back to the female plant and stimulates the development of the female sex cells, the oogonia.

What concerns us here is the chemotropism of the antheridial branches, as at present this is the only case of sexual chemotropism in the fungi or algae in which we known the identity of the attractant. T. C. McMorris (personal communication) has shown that the structural and stereochemical requirements for antheridiol hormone activity are highly specific. None of a wide range of synthetic isomers and related compounds had significant activity. The male *Achlya* cells efficiently remove added antheridiol and it cannot be extracted back from them (Barksdale, 1963b). Thus the sexual chemotropism of male cells of *Achlya* is elicited by a very specific steroid hormone which is inactivated by the recipient cells and which has a powerful morphogenetic action.

3. *Mucorales and trisporic acid*

Specific sexual chemotropism can be observed in the common mould, *Mucor*. The early stages of sexual reproduction in the Mucorales occur

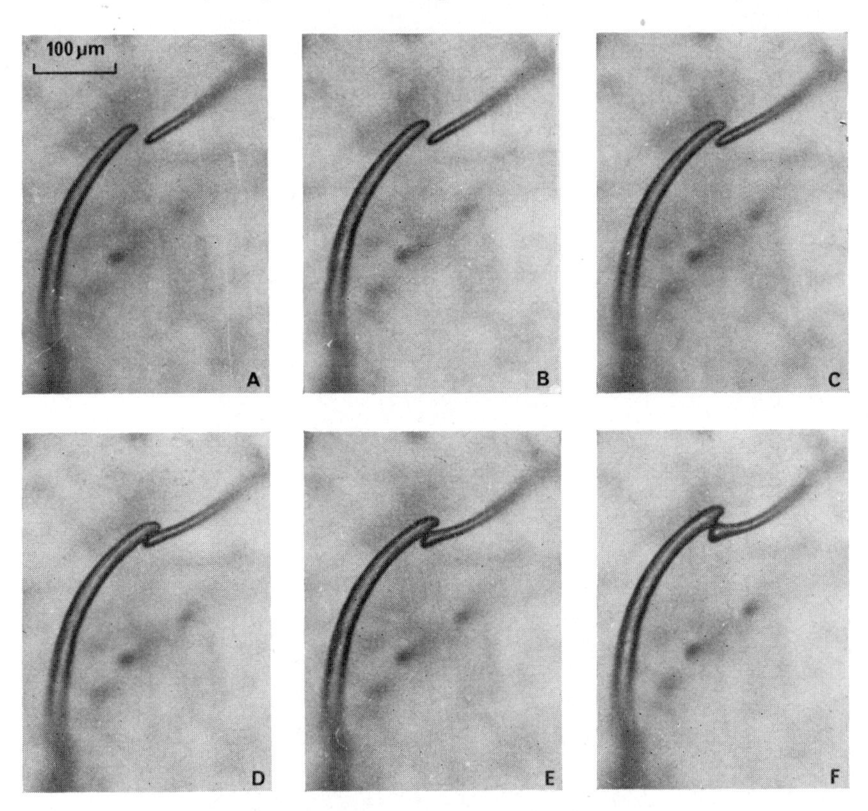

FIG. 6. Zygotropism just before contact between zygophores of *Mucor mucedo*, (+) on left, (—) on right. A, B, C, D, at 2 min intervals; E, F, at 4 min intervals. Cultures grown at 20°C on malt-glucose agar. Note asymmetry of zygophore tips just before meeting. Magnification × 110. (In part from Gooday, 1973.)

following the interaction of two mycelia of opposite mating type, (+) and (—), and involve the *de novo* synthesis of the sex hormone trisporic acid by both mating types and the response of both to trisporic acid by producing the characteristic sexual hyphae, the isomorphic zygophores (Gooday, 1973, 1974). Two zygophores of opposite mating type 2 mm apart will unerringly grow to meet and fuse; this is the phenomenon of 'zygotropism' (Fig. 6). Blakeslee (1904) stated that " . . . a mutual attraction which may be termed zygotactic, is exercised between the

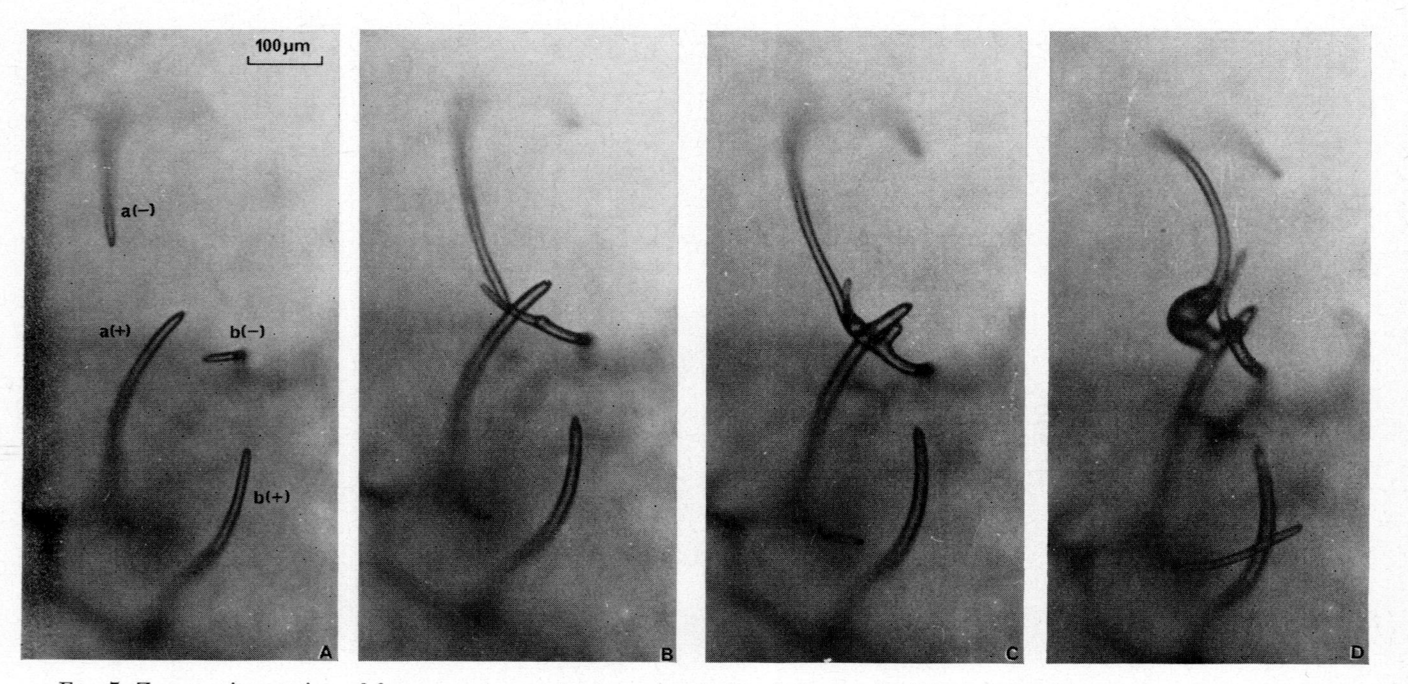

Fig. 7. Zygotropic meeting of four zygophores of *Mucor mucedo*. Conditions as for Fig. 5. (A) 3 Hours after meeting of the two cultures; (B) 35 min later; (C) 20 min later; (D) 1 hour later. The two zygophores, a(+) and a(—), fused successfully and formed a zygospore, the progametangia of which are seen in (D). The other two zygophores, b(+) and b(—), were unsuccessful, as neither fused and both stopped growing when a(+) and a(—) fused in (C). However, b(—) produced a peg, seen in (B), in response to the proximity of a(+) but before there was any contact between them.

zygophoric hyphae belonging to opposite mycelia and they may be seen to approach each other . . . Two minutes before contact occurred, and while the hyphae were separated by a distance equal to about a third of their width, very slight protrusions were observed on the sides mutually facing, seemingly as if the forces which were drawing the filaments laterally had effected a bulging of the delicate walls at their growing points." As well as this asymmetric bulging of zygophore tips (Fig. 6), the formation of a sub-apical peg is sometimes seen in *Mucor mucedo* (Fig. 7), which must be elicited by the zygotropic effectors, and is analogous to the peg formation during vegetative hyphal fusions discussed earlier. This is probably more important in some other species, such as *Mucor hiemalis*, where the ladder-like formation of several zygospores between the same two zygophores is seen much more often than in *M. mucedo*.

In *M. mucedo*, the zygophores grow steadily towards one another but cease elongating immediately they meet (Fig. 6, Table VI). If more

TABLE VI. Growth rate of two approaching zygophores of *Mucor mucedo*[a]

Time (min)	Growth rates (μm/min)	
	(+) zygophore	(—) zygophore
0–30	2·1	2·0
30–60	2·2	1·9
60–90	2·1	1·9
90–120	0	0

[a] Two zygophores approaching each other in a straight line. Measurements taken every two minutes. The zygophores fused at 90 min. The two cultures were grown on agar (malt extract, glucose, agar all 2% w/v) at 20°C in a moist chamber on the microscope stage and had met three hours previously.

complex interrelationships between several approaching zygophores are observed it is often seen that the unsuccessful zygophores also stop growing as soon as two zygophores meet (Fig. 7), as if a local source both of attraction and of growth stimulation is annulled as soon as any two cells fuse. This again is reminiscent of phenomena observed during vegetative hyphal fusions, when unsuccessful pegs cease growing (Fig. 3).

The special feature of this phenomenon for which there is no apparent biological parallel is that it is a specific mutual attraction through the air between two sexual cells that look identical in appearance and behaviour. Extensive early attempts to show a consistent observable sexual difference, such as size or pigmentation, between (+) and (—)

strains all met with failure (Gooday, 1973). The resulting necessity of envisaging a system with four sites of specificity (of production and of reception in each mating type) has proved a stumbling block in formulating theories as to the mechanism of zygotropism, and as discussed by Banbury (1955), the very existence of the phenomenon has often been denied or explained by chance meetings.

Banbury (1954, 1955) has provided description and photographic record of the zygotropism between zygophores of opposite mating type on adjacent blocks of agar, and showed how a pair will successfully meet through an out-of-line pinhole in a sheet of aluminium foil, but will fail to meet if an air current is passed between them. Banbury (1955) also suggests that there could be a repulsion between zygophores of opposite of mating type (likening the attraction between unlike cells and this repulsion between like cells to magnetic lines of force) but this has not been confirmed, and could represent a less specific negative autotropism rather than a sexual attribute of the zygophores. The zygophores do indeed grow away from their subtending vegetative mycelium, but are not positively phototropic or negatively geotropic as are their asexual counterparts, the sporangia. Plempel (1960, 1962, 1963) and Plempel and David (1961) have investigated the mechanisms of zygotropism and conclude that it is controlled by the production of two complementary gaseous growth substances, each produced by one mating type and active on the opposite mating type. They have shown that concomitantly with zygophore production there is in an increase in readily oxidizable volatile material from cultures of *M. mucedo* and ascribe this increase to the presence of the unstable zygotropic hormones, the concentration gradient of which would determine the orientation of zygophore growth. These authors also show that zygophores continue growing for a much longer time in the presence of vapours from compatible zygophores, and have a much reduced tendency to revert to sporangia.

Thus volatile chemicals must control zygotropism. In addition, there have been reports, notably by Burgeff (1924), that zygophore formation in *M. mucedo* can be stimulated by volatile agents from the opposite mating type. Other workers have failed to repeat this observation (Banbury, 1954; Plempel, 1963; Gooday, 1968), and Köhler (1935) genetically analysed Burgeff's strains of *M. mucedo* and showed that his results could be explained by heterokaryosis (see Gooday, 1973). It is now clear that trisporic acid, a C_{18} terpenoid derived from β-carotene, is the hormone stimulating zygophore production in both mating types and is produced by both mating types. Trisporic acid is not responsible for the volatile effect on zygophore production or growth, but it is uncertain how trisporic acid production itself is initiated in each mating type by agents from the opposite mating type. Gooday

(1973) has suggested that the as yet unidentified trisporate inducers could also act as zygotropic attractants just as antheridiol can induce hormone B production and also act as a chemotropic agent attracting the antheridia of *Achlya* (Barksdale, 1969).

Mesland *et al.* (1974) have now reported that they can confirm Burgeff's original observation of volatile induction of zygophore formation in *M. mucedo* by incubating two mats of vegetative mycelium of opposite mating type close to one another. These authors identify the active volatile agents with the neutral inducers of trisporic acid biosynthesis that Werkman and van den Ende (1973) have partially characterized from the two mating types of the related fungus *Blakeslea trispora*. Further, they report that the resulting zygophores are attracted by these agents and show the marked growth response characteristic of zygophores responding zygotropically. These zygophores will grow towards fresh vegetative mycelium of opposite mating type, and also towards a capillary tube or piece of filter paper containing a neutral extract of culture medium of opposite mating type. These authors suggest that vegetative mycelium of each mating type could produce metabolites with the dual effect on the opposite mating type of inducing trisporic acid biosynthesis and then of chemotropically attracting the resultant zygophores. Van den Ende (personal communication) tentatively identifies these metabolites from *M. mucedo* as 4-hydroxymethyl trisporate and trisporin (dimethyl precursor of trisporate) from (+) and (−) respectively. These are active on (−) and (+) respectively as zygotropic agents and trisporate precursors.

4. *Yeasts*

Yeasts can show sexually orientated growth responses. Sexual reproduction in many yeasts is controlled by the two alternative mating-type alleles, *a* and α, and haploid cells of opposite mating type conjugate to give diploid zygotes. Levi (1956) showed that a diffusible product could pass into the agar medium or through a cellophane membrane from α cells to cause the formation of copulatory processes by *a* cells and that these processes elongated towards the site of the α cells. Herman (1970a, b) studied the behaviour of adjacent cells of *Kluyveromyces* on agar. On a rich non-mating medium adjacent cells of the same or opposite mating type show a strong tendency to bud away from each other. This phenomenon corresponds to the negative autotropism shown during fungal spore germination. On a mating medium pairs of cells of identical mating type still budded away from each other, but sexually compatible pairs budded towards each other within two to three hours, so that eventually daughter cells came to lie next to each other. This effect could be seen even when the parent cells were 18–19 μm apart. It occurred between closely related species of *Kluyveromyces*, but not between more distantly related species (Herman, 1970b). These observations have been extended to other genera. *Hansenula anomala* and *S. cerevisiae* show an oriented budding by one mating type and an oriented cell enlargement by the other mating type. Intergeneric responses between opposite mating types of these species were observed (Herman, 1971). In the hormonal control of conjugation in *S. cerevisiae*, a peptide from α cells (Duntze *et al.*, 1970, 1973) and factors from both α and *a* cells (Yanagishima, 1969) are involved in the control of the cell enlargements and cessation of budding that are prerequisites for mating (reviewed by Gooday, 1974). It is, however, not clear if these compounds are directly involved in the zygotropic cell orientations observed by Levi and Herman.

Sexual conjugation also occurs between yeast-like cells of some Heterobasidiomycete fungi. In *Tremella mesenterica* the haploid basidiospores of the mating types "a" and "A" germinate and grow as yeast cells by budding. Bandoni (1963, 1965) has shown that cells of each mating type produce one or more diffusible hormones active over about 15 μm that initiate and then attract conjugation tubes from cells of opposite mating type. A similar mechanism has long been suggested for the conjugation of the sporidial yeast cells of the smut fungi, especially *Ustilago*. Poon *et al.* (1974) have directly observed mating in *Ustilago violaceae* and report that although most conjugations involved cells that were touching, a small number (4 out of 87) were between cells that were up to 4 μm apart. In these cases a peg developed on one of the cells only, and elongated to fuse with the other cell. This problem could be rewarding as there is now relevant genetical information on these fungi, e.g.: Cummings and Day (1973) have shown that conjugation in *U. violaceae* is limited to the G_1 phase of the cell cycle in the a_1 mating type, but can occur in most of the cell cycle in the a_2 mating type.

5. *Ascomycetes*

Sexual chemotropism in the Euascomycetes is well documented in several species and is possibly widespread. The best example is the work by Bistis (1956, 1957, 1965), Bistis and Raper (1963) and Bistis and Olive (1968) with the Discomycete, *Ascobolus stercorarius*. In this fungus the female ascogonium is a swollen hyphal branch which can develop a receptive cell, the trichogyne. The male cell can be a narrow hypha, the antheridium, or a spore-like oidium, and the trichogyne can fuse with either type of cell to allow plasmogamy. Karyogamy between the two nuclei is delayed until the young ascus develops. Bistis and his colleagues showed that the development of ascogonia and antheridia and the sexual activation of oidia are controlled by hormones diffusing between cells of the two mating types. Bistis (1956, 1957) demonstrated that a trichogyne will show clear directed growth towards an activated oidium. He micromanipulated the oidium to various new positions within 20 μm of the trichogyne, and found that the trichogyne changed direction to follow the oidium, even to the extent of pushing out new oriented branches. The cross-induction of sexual activation and differentiation between the two mating types is controlled by the mating-type alleles, A and a, but chemotropism is not, as a trichogyne will grow towards activated oidia of either mating type. In a natural environment the control of the initial sexual interactions would probably ensure that the trichogyne would be well placed to fuse with a compatible oidium. Following fusion there is a further developmental sequence probably involving chemotropism, as the ascogonium becomes ensheathed by neighbouring hyphae and new branches from its base which grow towards it.

Chemotropic attraction of the female trichogyne towards male sexual cells (spermatia) or male microconidia also occurs in *Neurospora sitophila* (Backus, 1939), *Bombardia lunata* (Zickler, 1937, 1952) and *Podospora anserina* (Esser, 1956). The spermatia do not require activation as do the oidia of *A. stercorarius*, and the attraction is controlled by the mating-type alleles. Zickler made chemotropically active cell-free extracts from *B. lunata* that specifically attracted trichogynes of the opposite mating type, and Esser obtained mutants of *P. anserina* lacking sexual chemotropism.

6. *Dikaryon and clamp connection formation in Basidiomycetes*

As discussed earlier, vegetative hyphal fusions are common in Basidiomycetes. The fusions can have sexual significance if the fusing cells are heteroallelic for incompatibility factors. Plasmogamy can occur between compatible cells at any time during growth, but instead of being quickly followed by karyogamy, it results in the formation of a dikaryotic mycelium in which the two compatible nuclei divide synchronously so that each cell in the mycelium contains the same two compatible nuclei. In most species this apportioning of pairs of nuclei is physically associated with the formation of clamp connections, in which one nucleus divides through a short hyphal branch which then arches over backwards and fuses with the main hypha in which the other nucleus has meanwhile divided. Karyogamy is delayed until, perhaps after many years, a fruit body is formed and the two nuclei fuse in the basidium on the gills just before meiosis. Buller (1931, 1933) discussed the hyphal fusions that can lead to dikaryotization and concluded that they have no sexual significance, for fusions are common in these fungi between cells of like or unlike mating type, and fusions often increase in number after dikaryotization. However, Ahmad and Miles (1970) have concluded that the "A" incompatibility factor has a role in controlling hyphal fusions in *Schizophyllum commune*. By making counts of the frequencies of hyphal fusions they have shown that hyphae with different A factors fuse more often than hyphae with the same A factor. This fungus is tetrapolar, with fruiting normally only occurring in dikaryons heteroallelic for the two incompatibility factors ($A \neq B \neq$). Calculations from their data for counts of about 500 cell-cell contacts show that in compatible ($A \neq B \neq$) and common B ($A \neq B =$) contacts 2·9% and 3·1% of the hyphae

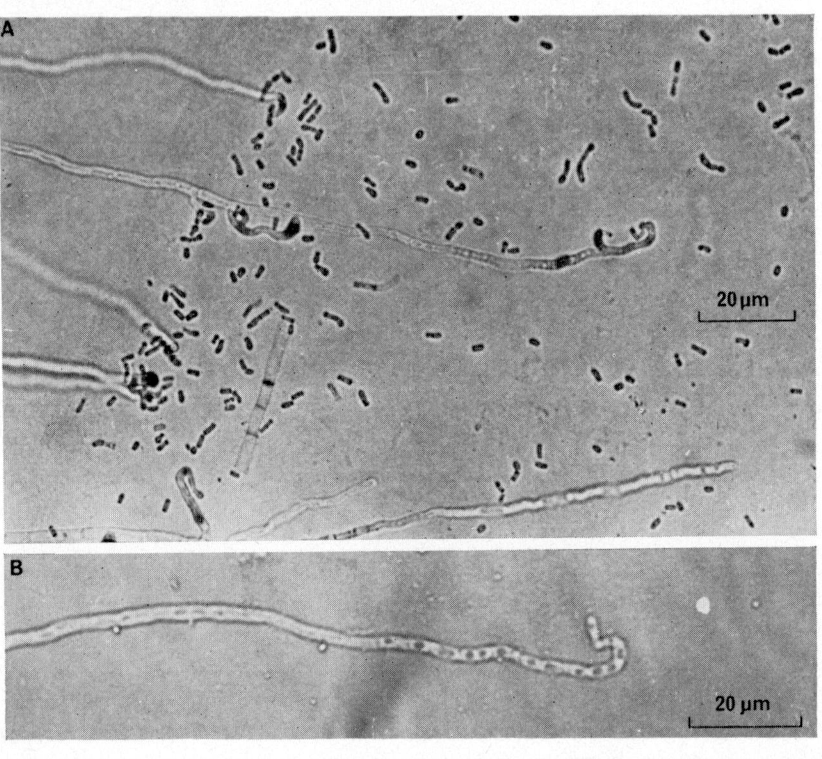

FIG. 8. Homing of hyphal tips to oidia of different isolates of *Psathryella* species growing at 25°C. The oidia had been smeared in front of the hyphae 2 hours earlier, and the homing and fusion have been followed by lethal reactions, shown as a vacuolation in (B). (A) × 670; (B) × 770. Photographs provided by Dr. R. F. O. Kemp.

fused, but between common A ($A = B \neq$) only 0·7%. In compatible matings, clear hyphal attractions were observed when the cells were within 10 μm. As discussed earlier, hyphal fusions must be mediated by diffusible materials, and Ahmad and Miles now have provided evidence for this. One strain was grown on agar under a cellophane membrane, on which was inoculated an "inducing strain". After incubation for 6 days the lower strain was faced with a testing strain and the hyphal fusions scored. Mycelia that had been in diffusion contact with sexually compatible cells showed statistically more hyphal fusions than those that had been in diffusion contact with mycelia of the same mating type. For example, the fusion frequency for the compatible cross 699 × 3054 rose from 3·8% to 9·4% when 699 and 3054 had been "induced" by 3054 and 699 respectively, and the fusion frequencies for self-meetings of 699 × 699 and 3054 × 3054 rose from 1·4% and 0·4% to 6·2% and 6·4% when the mycelia had been "induced" by compatible cells. Thus heterozygosity at A leads to a high fusion frequency that appears to be controlled by diffusible chemotropic factors.

In many Basidiomycetes dikaryotization can also occur following plasmogamy between a vegetative hypha and an oidium of compatible mating type. Oidia are characteristically uninucleate spore-like cells produced in clusters by monokaryotic mycelium. The "homing" of vegetative hyphae towards oidia has been investigated in *Coprinus*, *Psathyrella* and *Flammulina* species (Kemp, 1970, personal communication; Jurand and Kemp, 1972) and in *Clitocybe truncicola* (Bistis, 1970). In *Coprinus* species oidia can stimulate hyphae to grow towards them on agar from distances exceeding 75 μm and homing can be seen 15 minutes after adding oidia to the agar around the hyphal tips (Fig. 8). Although there are no published quantitative data it would seem that the oidial to hyphal stimulus results in a higher frequency of successful fusions than a hyphal-hyphal interaction. Chemotropism must be the mechanism of attraction, so the oidium may be a richer source of the chemotropic factor than a vegetative hyphal tip. Just as with hyphal-hyphal fusions, oidial homing is shown by monokaryotic and dikaryotic hyphae regardless of the incompatibility factors of the oidium and hypha. Data are not available to show whether it is more efficient between cells of different mating type, as with the hyphal fusions of *S. commune* discussed above, and so it is not clear to what extent it is controlled by incompatibility factors. Homing is not totally species specific in *Coprinus* and *Psathyrella* (Jurand and Kemp, 1972; R.F.O. Kemp, personal communication), although consequent hybrid plasmogamy terminates in a lethal reaction (Fig. 8).

It is not clear whether or not there is a sexual significance in the formation of clamp connections, for, as with plasmogamy leading to dikaryotization, there is a close parallel between the fusion of the tip of the hook cell and the main hypha and the phenomenon of vegetative fusions between adjacent hyphae of the same species. Buller (1933) discusses clamp connection formation as a 'hook to peg' fusion, and emphasises: "Investigations on the details of clamp-connexion formation of Coprinus have taught me that the hook of a clamp-connexion does not fuse directly with the main hypha as hitherto has been supposed, but with a blunt process or peg sent out by the main hypha in response to a stimulus given by the apex of the hook" (Fig. 9). This peg can clearly be seen in the photographs of Niederpruem *et al.* (1971). Harder

(1927) had earlier suggested that chemotropic hormones could be involved in controlling the turning and re-fusion of the hook cell but there is no evidence for this. There is considerable evidence for the genetic control of clamp connection formation, particularly in *S. commune* (reviewed by Raper, 1966; Burnett, 1968; Niederpruem and Wessels, 1969). Normal clamp connections are formed in dikaryons heteroallelic for the two incompatibility factors ($A \neq B \neq$); false clamp connections (without hook cell fusion) are formed in common B heterokaryons ($A \neq B =$); but no clamp connections are formed in common A heterokaryons ($A = B \neq$). Thus the A locus appears to control clamp connection formation, but the fusion itself also involves the B locus. As described above, the A locus also plays a part in controlling cell-cell fusions in this fungus. The B factor and to a much lesser extent the A

FIG. 9. Stages in the hook-to-peg fusion during the formation of a clamp connection in *Coprinus*. Note the formation of the hyphal peg in response to the hook cell. In *C. lagopus* the hyphal diameter was about 2·5 μm, and the drawings represent 3, 15, 21, 22, and 23 minutes after the time at which the hook cell started to be formed. After Buller (1933).

factor have been shown to control the production of an important morphogenetic enzyme, the wall lytic R-glucanase (Wessels and Niederpruem, 1967), which is perhaps involved in apical hyphal extension and chemotropic bending and certainly in cell fusion.

The relationships between these three phenomena—the hyphal fusions of compatible mating types, homing, and hook cell fusion—and vegetative hyphal fusion remains unclear. Perhaps all represent the same phenomenon modulated by interactions between two compatible mating types, or perhaps there are two mechanisms, one vegetative and one sexual, and the final result is determined by a summation of the two. The results of Ahmad and Miles (1970) do not differentiate between these two possibilities, but further characterization of their diffusible factor would help to do so.

C. *The specificity of sex attractants*

The fungal and algal sex attractants that have been partially or completely characterized are in the category of *secondary metabolites* produced

Table VII. Properties of fungal and algal sex attractants

	Antheridiol	Sirenin	Ectocarpen	Fucoserraten	Multifidene
Formula, mol wt	$C_{20}H_{42}O_5$, 470	$C_{15}H_{24}O_2$, 236	$C_{11}H_{16}$, 148	C_8H_{12}, 108	$C_{11}H_{16}$, 148
Probable precursor	Cholesterol	Farnesyl pyrophosphate	C_{10} polyene, or linolenic acid?	—	C_{11} polyene alcohol
Specificity of production and activity	*Achlya* spp.	*Allomyces* spp.	*Ectocarpus* spp.	*Fucus* spp.	*Cutleria* spp.
Site and control of synthesis	♀ Cells	♀ Gametes	♀ Gametes	Oogonia and eggs	♀ Macrogametes
Optimal yield	6×10^{-9} Molar in culture filtrate	10^{-6} Molar in culture filtrate	80 μg from 1 g algae	2·7 ng from 1 g algae	2·5 μg from 1 g algae
Sensitivity of bioassay (M)	10^{-11} (antheridal induction)	5×10^{-11}	—	—	—
Attractive action	Chemotropism of antheridia	Chemotaxis of gametes	Chemotaxis of gametes	Chemotaxis of gametes	Chemotaxis of ♂ microgametes
Stability in culture	Inactivated by ♂ cells	Inactivated by ♂ cells	Volatile	Volatile	Volatile
References	Barksdale, 1969; Arsenault *et al.*, 1968	Machlis, 1974a, b; Machlis *et al.*, 1966	Müller *et al.*, 1971; Jaenicke & Müller, 1973	Müller, 1972b; Müller & Jaenicke, 1973	Müller, 1974, Jaenicke *et al.*, 1974.

after the exponential growth phase. Secondary metabolites are often associated with differentiation but few have any apparent function. Bu'Lock (1967) stated that "secondary metabolites are an expression of the individuality of the species in molecular terms, and possibly these are the most appropriate terms in which that individuality can be expressed, since the 'senses' of micro-organisms are primarily chemical. By their fruits they shall be known . . . but the mycelium has no need of a microscope". Nowhere is this more relevant than with these sex

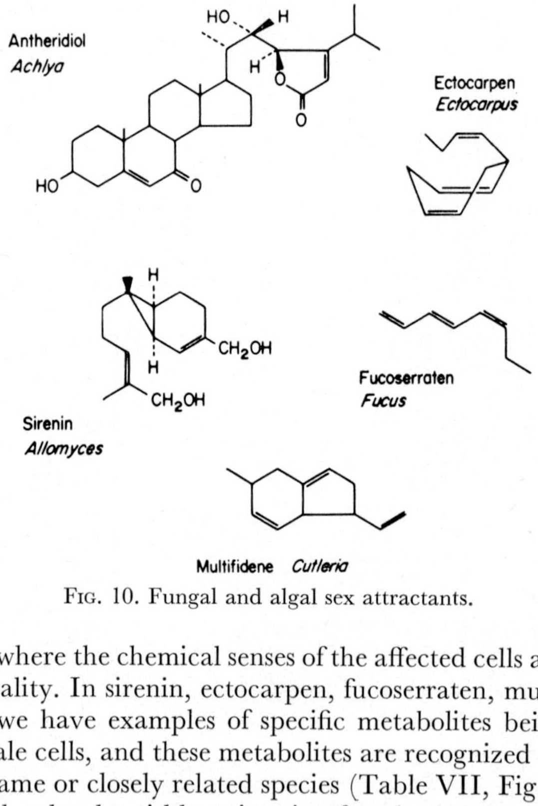

FIG. 10. Fungal and algal sex attractants.

attractants, where the chemical senses of the affected cells are a function of their sexuality. In sirenin, ectocarpen, fucoserraten, multifidene and antheridiol we have examples of specific metabolites being produced only by female cells, and these metabolites are recognized only by male cells of the same or closely related species (Table VII, Fig. 10). Sirenin is produced by the sluggishly swimming female eamete, ectocarpen by the female gamete only when it has settled, fucoserraten by oogonia and eggs. Antheridiol and the *Mucor* zygophore attractants appear to have a dual role by initiating sexual differentiation in a receptive compatible partner, in the case of *Mucor* via trisporic acid, and then acting as attractants of the resultant sexual hyphae. Only in the Basidiomycetes and to some extent the Ascomycetes does the tight sexual specificity of production and action of attractants seem to be lacking. This perhaps reflects the progressive loss of differentiated sexual organs in the higher fungi.

VI. Discussion

A. *Mechanisms of chemotropism*

A discussion of chemotropism in fungi requires knowledge of the way in which hyphae grow. This was realized long ago by Ward (1888) and Reinhardt (1892) who discussed hyphal growth in the context of hyphal fusion. Ward's view of the mode of growth of a hypha coincides with the popular current view (except that it is now known that cellulose is not a component of *Botrytis*) as, after suggesting that branching and peg formation are controlled by "a ferment capable of swelling and dissolving cellulose", he goes on: "I imagine, moreover, that the continuous forward growth of the apex of any hypha takes place in a similar way, that is to say, the ferment-substance at the apex keeps the cellulose of the hypha at that place in a soft, extensible condition, and the pressure from behind stretches it and drives the tip forward." This suggestion contains the essence of the integrated model of apical growth in fungi put forward by Bartnicki-Garcia (1973) at the conclusion of a review of hyphal morphogenesis. Reinhardt (1892) elegantly showed that hyphal growth is confined to the extreme apex by watching whether or not small particles were displaced when sprinkled on the hypha. Autoradiographic studies have now confirmed that the extreme apex is the major site of deposition of cell wall polymers such as chitin and glucans (Bartnicki-Garcia and Lippman, 1969; Gooday, 1971). Microscopic examination of a growing hypha shows that the extreme apex is devoid of recognizable organelles, and in many fungi contains a densely staining body, the "Spitzenkörper". In the electron microscope this area is seen to have a characteristic collection of vesicles which show every appearance of having been derived from the sub-apical endomembrane system, and of eventually fusing with the apical cell membrane (Girbardt, 1969; Grove and Bracker, 1970; Grove *et al.*, 1970). It is highly probable that these vesicles represent the synthetic and the lytic activities necessary for orderly apical growth. Behind these vesicles is a dense array of mitochondria that with phase-contrast or interference microscopy can be seen jostling about. Nuclei and vacuoles are then found behind this mitochondrial zone.

An important consequence of the model of hyphal growth involving controlled lysis and synthesis is that growth stimulation on one side of the apex will displace the apical lysis and synthesis to that side and so cause a bending towards the stimulator. Growth inhibition will cause a bending away. This situation is the opposite in cause and effect to the phenomenon familiar to plant physiologists of higher plant apices bending away from a growth stimulator by intercalary growth.

A chemotropic response must involve the recognition of an asymmetric distribution of the chemical at the growing apex itself to result in the observed oriented growth. Simple calculations show that the gradients across the cell must be very small. The range of chemical effectors is wide; oxygen or a staling factor in the negative autotropism of a variety of fungi, a putative macromolecular species-specific chemical in the positive autotropism of *Botrytis* and of hyphal fusions in the higher fungi, mixtures of amino acids for *Saprolegnia*, antheridiol for antheridial hyphae of *Achlya*, two sex specific volatile effectors for zygophores of *Mucor*, and specific sex sttractants in Ascomycetes and Basidiomycetes. These chemical effectors are so diverse that it seems unlikely that they all have the same cellular site of action. For oxygen as an attractant the aggregation of mitochondria just behind the tip is a prime contender as the receptor site. Mitochondria on the oxygen-rich side of the hypha would respire faster than those on the oxygen-poor side. A faster rate of metabolism could result in asymmetric growth and bulging in favour of the oxygen-rich side, and so the apex would turn up the oxygen gradient. If the *Botrytis* attractant and the vegetative fusion attractant are macromolecules, then the cell surface, either the growing "metabolic" wall or the growing cell membrane, is the most likely site of action. The sex attractants are small specific hormone molecules and a mechanism involving specific binding to receptor molecules in the apical cell membrane is plausible. The search for such a receptor molecule should be possible with the successful preparation and characterization of the hormones themselves. It is already known that antheridiol is irreversibly removed from solution by male cells of Achlya (Barksdale, 1967, 1969).

There are suggestions of bioelectric control of the polarity of hyphal growth (Jaffe, 1968; Bartnicki-Garcia, 1973), but it is premature to try to interpret chemotropism by such a mechanism.

It must be remembered that most or all of the chemical attractants also act as growth regulators. They can control the initiation, the site of initiation, and the sustained growth of hyphae. The negative autotropic factors control the site of germ tube origin, and possibly the site and to some extent the frequency of secondary and lower order space-filling branching. The positive autotropic factors control the site of germ tube origin in *Botrytis*, and initiate peg formation at specific sites. Increasing concentrations of antheridiol can elicit a range of sexual morphogenetic responses from male *Achlya* hyphae. Low concentrations initiate the formation of antheridial branches, and higher concentrations attract these branches and sustain their growth for longer periods (Barksdale, 1969). Zygophores of *Mucor* responding to their attractants grow much longer than others, and pegs are initiated just before contact of compatible cells. The zygophores themselves are initiated by trisporic acid, and Mesland *et al.* (1974) identify the zygophore attractants as inducers and perhaps also precursors of trisporic acid.

Thus the chemotropic attractants have a more profound relationship with the cell apex than just through eliciting asymmetric growth with an asymmetric concentration. The initiation of a hyphal branch must involve a localized reaggregation or differentiation of cell components, so that the wall is weakened, Spitzenkörper vesicles formed, and new wall synthesis induced. The presence of the effector can then maintain the integrity of this new hyphal apex.

B. *Mechanisms of chemotaxis*

The chemotaxes described above have all involved flagellated cells. Details of the fine structure of most of these cells and of their flagella are now recorded (see reviews by Kole (1965) for fungi and Manton (1970) for plant spermatozoids) but these studies do not at present directly point to possible mechanisms for chemotaxis. The cells vary from the uniflagellate *Allomyces* spores to the *Oedogonium* spermatozoa and zoospores, which have an elaborate ring of about 30 and 120 flagella respectively. Müller and Falk (1973) have examined male and female gametes of *Ectocarpus* in detail, but although these cells differ in behaviour both by responding to and producing ectocarpen and during fertilization, they are identical in appearance. The chemotactic responses of most of these cells are limited to one or a few effectors. The gametes are the most specific, responding to attractants such as sirenin, ectocarpen and fucoserraten, while the zoospores respond to metabolites such as amino acids, carbohydrates and ethanol. Once the chemotactic effectors have been identified, it should prove possible to search for possible receptor molecules by extracting materials with the property of specifically binding the effectors. Current techniques such as affinity chromatography could be utilized, and the distributions of these putative binding molecules could be correlated with the known chemotactic responses of different cells.

Where are the chemoreceptors: in the cell, on the cell, or on the flagella? For the chemotaxes described here the cell has to quickly and specifically respond to the concentration gradient across itself, and so as suggested by Carlile (1966) the cell surface is the most likely site. The chemoreceptors would be arranged around the body of the cell itself. The flagella are also enclosed in the cell membrane, and if they were chemoreceptive the cell would be able to span a larger concentration gradient. However, more critical work is needed to determine whether these cells can always sense the concentration gradient across themselves directly, or whether they can monitor any change in external concentration with time.

Experiments with protozoa suggest that the state of the cell membrane is of prime importance in the control of ciliary beating (Naitoh and Eckert,

1969a, b; Eckert, 1972; Naitoh and Kaneko, 1972). Using electrodes in single cells these authors showed that when the anterior of *Paramecium* was mechanically stimulated there was a transient increase in membrane permeability to calcium ions, and the consequent influx of calcium ions led to a transient depolarization of the cell, which in turn was followed by a reverse of direction of ciliary beating. When the posterior of the cell was mechanically stimulated, an increased permeability to potassium ions, an efflux of potassium ions, hyperpolarization of the cell, and an increase in ciliary beating in the normal direction were observed. Similarly, a depolarization of a cell of *Euplotes* resulted in the reversing of direction of the cirri, while a hyperpolarization resulted in the normal forward orientation. This orientation response of the cirri was independent of the "neuromotor" fibrils in this cell, as surgical cutting of these made no difference. Thus these authors suggest that the state of polarization of the cell has a controlling influence on the direction and speed of swimming of a ciliate. Most of the chemotactic fungal and algal cells are too small for conventional microelectrode techniques but their state of polarization in the presence and absence of chemotactic effectors could be deduced indirectly by measuring ionic fluxes in different conditions. That specific chemicals can change the polarization of receptive cells is known for animal taste bud cells, which show a depolarization on chemical stimulation (Beidler, 1971). There is also indirect evidence that glutamate, an attractant for some zoospores, can act as an excitatory neurotransmitter, causing specific depolarizations at synapses (Snyder *et al.*, 1973). It has been suggested that amoeboid movement, and hence amoeboid chemotaxis, is controlled by localized differences in membrane potential (Wolpert and Gingell, 1968). These authors also suggest that a model involving effects on the fixed surface potential, as opposed to the transmembrane potential, provides a worthwhile alternative with which to interpret the experimental observations.

Acknowledgments

I thank the many authors who have sent me reprints and preprints of their work, and Drs R. F. O. Kemp, I. C. Harvey and D. G. Müller for photographs.

References

Ahmad, S. S. and Miles, P. G. (1970a). *Genet. Res.* **15,** 19–28.

Ahmad, S. S. and Miles, P. G. (1970b). *Mycologia* **62,** 1008–1017.

Allen, R. N. (1973). "Aspects of Chemotaxis in Zoospores of *Phytophthora cinnamomi*." Ph.D. Thesis, University of Auckland.

Allen, R. N. and Harvey, J. D. (1974). *J. Gen. Microbiol.* **84,** 28–38.

Allen, R. N. and Newhook, F. J. (1973). *Trans. Br. Mycol. Soc.* **61,** 287–302.

Allen, R. N. and Newhook, F. J. (1974). *Trans. Br. Mycol. Soc.* **63,** 383–385.
Arens, K. (1929). *Jahrb. Wiss. Bot.* **70,** 93–157.
Arsenault, G. P., Biemann, K., Barksdale, A. W. and McMorris, T. C. (1968). *J. Am. Chem. Soc.* **90,** 5635–5636.
Backus, M. P. (1939). *Bull. Torrey Bot. Club* **66,** 63–76.
Bald, J. G. (1952). *Am. J. Bot.* **39,** 97–99.
Banbury, G. H. (1954). *Nature* **173,** 499.
Banbury, G. H. (1955). *J. Exp. Bot.* **6,** 235–244.
Bandoni, R. J. (1963). *Can. J. Bot.* **41,** 467–474.
Bandoni, R. J. (1965). *Can. J. Bot.* **43,** 627–630.
Baracchini, O. and Sherris, J. C. (1959). *J. Pathol. Bact.* **77,** 565–574.
Barksdale, A. W. (1963a). *Mycologia* **55,** 627–632.
Barksdale, A. W. (1963b). *Mycologia* **55,** 164–171.
Barksdale, A. W. (1967). *Ann. N.Y. Acad. Sci.* **144,** 313–319.
Barksdale, A. W. (1969). *Science* **166,** 831–837.
Barnett, H. L. and Binder, F. L. (1973). *Ann. Rev. Phytopath.* **11,** 273–292.
Bartnicki-Garcia, S. (1973). *In* "Microbial Differentiation" (Ashworth, J. M. and Smith, J. E. ed.), pp. 245–267. 23rd Symposium of the Society for General Microbiology. Cambridge University Press.
Bartnicki-Garcia, S. and Lippman, E. (1969). *Science* **165,** 302–304.
Beidler, L. M. (1971). *In* "Handbook of Sensory Physiology" (L. M. Beidler, ed.), Vol. IV, part 2, pp. 200–220. Springer, Berlin.
Berry, C. R. and Barnett, H. L. (1957). *Mycologia* **49,** 374–386.
Bhalerao, U. T., Plattner, J. J. and Rapoport, H. (1970). *J. Am. Chem. Soc.* **92,** 3429–3433.
Bistis, G. N. (1956). *Am. J. Bot.* **43,** 389–394.
Bistis, G. N. (1957). *Am. J. Bot.* **44,** 436–443.
Bistis, G. N. (1965). *In* "Incompatibility in Fungi" (K. Esser and J. R. Raper, eds.), pp. 23–31. Springer, Berlin.
Bistis, G. N. (1970). *Mycologia* **62,** 911–923.
Bistis, G. N. and Olive, C. S. (1968). *Am. J. Bot.* **55,** 629–634.
Bistis, G. N. and Raper, J. R. (1963). *Am. J. Bot.* **50,** 880–891.
Blakeslee, A. F. (1904). *Proc. Am. Acad. Arts Sci.* **40,** 205–319.
Bonner, J. T. (1970). *In* "Chemical Ecology" (E. Sondheimer and J. B. Simeone, eds.), pp. 1–19. Academic Press, New York.
Brown, R. M., Johnson, C. and Bold, H. C. (1968). *J. Phycol.* **4,** 100–120.
Buller, A. H. R. (1931). "Researches on Fungi", Vol. 4. Longmans Green, London.
Buller, A. H. R. (1933). "Researches on Fungi", Vol. 5. Longmans Green, London.
Bu'Lock, J. D. (1967). "Essays in Biosynthesis and Microbial Development". Wiley, New York.
Burgeff, H. (1924). *Bot. Abh.* **4,** 1–135.
Burnett, J. H. (1968). "Fundamentals of Mycology". Arnold, London.
Carlile, M. J. (1966). *In* "The Fungus Spore" (M. F. Madelin, ed.), pp. 175–186. Butterworths, London.
Carlile, M. J., Dickens, J. S. W., Mordue, E. M. and Schipper, M. A. (1962). *Trans. Br. Mycol. Soc.* **45,** 457–461.

CARLILE, M. J. and MACHLIS, L. (1965a). *Am. J. Bot.* **52,** 478–483.
CARLILE, M. J. and MACHLIS, L. (1965b). *Am. J. Bot.* **52,** 484–486.
CHANG-HO, Y. and HICKMAN, C. J. (1970). *In* "Root Diseases and Soil-borne Pathogens" (T. A. Toussoun, R. V. Bega and P. E. Nelson, eds.), pp. 109–111. Univ. Calif. Press, Berkeley.
CLARK, J. F. (1902). *Bot. Gaz.* **33**, 26–48.
COOK, A. H. and ELVIDGE, J. A. (1951). *Proc. R. Soc. Lond.* **138B**, 97–114.
COOK, A. H., ELVIDGE, J. A. and HEILBRON, I. (1948). *Proc. R. Soc. Lond.* **135B,** 293–301.
CUMMINS, J. E. and DAY, A. W. (1973). *Nature* **245,** 259–260.
DILL, B. C. and FULLER, M. S. (1971). *Arch. Mikrobiol.* **78,** 92–98.
DUKES, P. D. and APPLE, J. L. (1961). *Phytopathology* **51,** 195–197.
DUNTZE, W., MACKAY, V. and MANNEY, T. R. (1970). *Science* **168,** 1472–1473.
DUNTZE, W., STÖTZLER, D., BÜCKING-THROM, E. and KALBITZER, S. (1973). *Eur. J. Biochem.* **35,** 357–365.
ECKERT, R. (1972). *Science* **176,** 473–481.
EMERSON, R. and WILSON, C. M. (1954). *Mycologia* **46,** 393–434.
ESSER, K. (1956). *Z. Indukt. Abstammungs-Vererbungsl.* **87,** 595–624.
FISCHER, F. G. and WERNER, G. (1955). *Hoppe-Seyler's Z. Physiol. Chem.* **300,** 211–236.
FISCHER, F. G. and WERNER, G. (1958). *Hoppe-Seyler's Z. Physiol. Chem.* **310,** 65–91.
FLENTJE, N. T. (1959). *In* "Plant Pathology. Problems and Progress, 1908–1958" (C. S. Holton *et al.*, eds.), pp. 76–87. University of Wisconsin Press, Wisconsin.
FULTON, H. R. (1906). *Bot. Gaz.* **41,** 81–108.
GIRBARDT, M. (1969). *Protoplasma* **67,** 413–441.
GOODAY, G. W. (1968). *New Phytol.* **67,** 815–821.
GOODAY, G. W. (1970). *J. Mar. Biol. Assoc. U.K.* **50,** 199–208.
GOODAY, G. W. (1971). *J. Gen. Microbiol.* **67,** 125–133.
GOODAY, G. W. (1972). *Biochem. J.* **127,** 2–3p.
GOODAY, G. W. (1973). *In* "Microbial Differentiation" (Ashworth, J. M. and Smith, J. E., eds.), pp. 269–294. 23rd Symposium of the Society for General Microbiology. Cambridge University Press.
GOODAY, G. W. (1974). *Ann. Rev. Biochem.* **43,** 35–49.
GOODE, P. M. (1956). *Trans. Br. Mycol. Soc.* **39,** 367–377.
GRAVES, A. H. (1916). *Bot. Gaz.* **62,** 337–369.
GREGORY, C. T. (1912). *Phytopathology* **2,** 235–249.
GROVE, S. N. and BRACKER, C. E. (1970). *J. Bact.* **104,** 989–1009.
GROVE, S. N., BRACKER, C. E. and MORRE, D. J. (1970). *Am. J. Bot.* **59,** 245–266.
HARDER, R. B. (1927). *Zeit. Bot.* **19,** 337–407.
HARVEY, I. C. (1975). *Trans. Br. Mycol. Soc.* **64,** 489–495.
HERMAN, A. I. (1970a). *Can. J. Gen. Cytol.* **12,** 66–69.
HERMAN, A. I. (1970b). *Antonie van Leeuwenhoek J. Microbiol. Serol.* **36,** 421–425.
HERMAN, A. I. (1971). *Antonie van Leeuwenhoek J. Microbiol. Serol.* **37,** 379–384.
HICKMAN, C. J. (1970). *Phytopathology* **60,** 1128–1135.
HICKMAN, C. J. and HO, H. H. (1966). *Ann. Rev. Phytopathol.* **4,** 195–200.

Hoffman, L. R. (1960). *Southwestern Naturalist* **5,** 111–116.
Hoffman, L. R. (1961). "Studies of the Morphology, Cytology, and Reproduction of *Oedogonium* and *Oedocladium*." Dissertation, Univ. of Texas, Austin, Texas.
Holligan, P. M. and Gooday, G. W. (1975). *Symp. Soc. Exp. Biol.* **29,** 205–227.
Jaenicke, L. (1972a). *Biochem. J.* **128,** 30p.
Jaenicke, L. (1972b). "Sexuallockstoffe in Pflanzenreich". Rheinisch-Westfülische Akademie der Wissenschaften. Vorträge No. 217, 48pp.
Jaenicke, L., Akintobi, T. and Müller, D. G. (1971). *Angew. Chem. Int. Ed. Engl.* **10,** 492–493.
Jaenicke, L. and Müller, D. G. (1973). *Fortschr. Chem. Org. Naturst.* **30,** 61–100.
Jaenicke, L., Müller, D. G. and Moore, R. E. (1974). *J. Am. Chem. Soc.* **96,** 3324–3325.
Jaffe, L. F. (1965). *Biophys. J.* **5,** 201–210.
Jaffe, L. F. (1966). *Plant Physiol.* **41,** 303–306.
Jaffe, L. F. (1968). *Adv. Morphog.* **7,** 295–328.
Jurand, M. K. and Kemp, R. F. O. (1972). *J. Gen. Microbiol.* **72,** 575–579.
Katsura, K., Masago, H. and Miyata, Y. (1966). *Ann. Phytopathol. Soc. Jap.* **32,** 215–220.
Katsura, K. and Miyata, Y. (1971). *In* "Morphological and Biochemical Events in Plant-Parasite Interaction" (S. Akai, and S. Ouchi, eds.), pp. 107–128, Phytopathol. Soc. Japan, Tokyo.
Keeble, F. (1910). "Plant-Animals—A Study in Symbiosis". Cambridge University Press, Cambridge.
Keeble, F. and Gamble, F. W. (1907). *Q. J. Microsc. Sci.* **51,** 167–219.
Khew, K. L. and Zentmeyer, G. A. (1973). *Phytopathology* **63,** 1511–1517.
Kemp, R. F. O. (1970). *Trans. Brit. Mycol. Soc.* **54,** 488–489.
Köhler, E. (1929). *Planta* **8,** 140–153.
Köhler, E. (1930). *Planta* **10,** 495–522.
Köhler, F. (1935). *Z. Indukt. Abstammungs-Verebungsl.* **70,** 1–54.
Köhler, K. (1967). *In* "Handbuch der Pflanzenphysiologie" (W. Ruhland, ed.), Vol. 18, pp. 282–320. Springer, Berlin.
Kole, A. P. (1966). *In* "The Fungi" (G. C. Ainsworth and A. S. Sussman, eds.), Vol. 1, pp. 86–91. Academic Press, New York.
Levi, J. D. (1956). *Nature* **177,** 753–754.
McMorris, T. C. and Barksdale, A. W. (1967). *Nature* **215,** 320–321.
Machlis, L. (1958a). *Nature* **181,** 1790–1791.
Machlis, L. (1958b). *Physiol. Plant.* **11,** 181–192.
Machlis, L. (1958c). *Physiol. Plant.* **11,** 845–854.
Machlis, L. (1968). *Plant Physiol.* **43,** 1319–1320.
Machlis, L. (1969a). *Physiol. Plant.* **22,** 126–139.
Machlis, L. (1969b). *Physiol. Plant.* **22,** 392–400.
Machlis, L. (1973a). *Plant Physiol.* **52,** 524–526.
Machlis, L. (1973b). *Plant Physiol.* **52,** 527–530.
Machlis, L., Hill, G. G. C., Steinback, K. E. and Reed, W. (1974). *Physiol. Plant.*

MACHLIS, L., NUTTING, W. H. and RAPOPORT, H. (1968). *J. Am. Chem. Soc.* **90,** 1674–1676.
MACHLIS, L., NUTTING, W. H., WILLIAMS, M. W. and RAPOPORT, H. (1966). *Biochemistry* **5,** 2147–2152.
MACHLIS, L. and RAWITSCHER-KUNKEL, E. (1963). *Int. Rev. Cytol.* **15,** 97–138.
MACHLIS, L. and RAWITSCHER-KUNKEL, E. (1967). *In* "Fertilization" (C. B. Metz and A. Monroy, eds.), Vol. 1, pp. 117–161. Academic Press, New York.
MANTON, I. (1970). *In* "Comparative Spermatology" (B. Baccetti, ed.), pp. 143–158. Academic Press, New York.
MASSEE, G. (1905). *Philos. Trans. Roy. Soc. B.* **197,** 7–24.
MEHROTRA, R. S. (1970). *Can. J. Bot.* **48,** 879–882.
MESLAND, D. A. M., HUISMAN, J. G., and ENDE, H. VAN DEN (1974). *J. Gen. Microbiol.* **80,** 111–117.
MIYOSHI, M. (1894). *Bot. Zeit.* **52,** 1–28.
MÜLLER, D. and JAFFE, L. F. (1965). *Biophys. J.* **5,** 317–335.
MÜLLER, D. G. (1967a). *Planta* **75,** 39–54.
MÜLLER, D. G. (1967b). *Naturwissenschaften* **54,** 496–497.
MÜLLER, D. G. (1968). *Planta* **81,** 160–168.
MÜLLER, D. G. (1969). *Naturwissenschaften* **56,** 220.
MÜLLER, D. G. (1972a). *Bull. Soc. Bot. Fr. Mém.*, 87–98.
MÜLLER, D. G. (1972b). *Naturwissenschaften* **59,** 166.
MÜLLER, D. G. (1974). *Biochem. Physiol. Pflanzen* **165,** 212–215.
MÜLLER, D. G. and FALK, H. (1973). *Arch. Microbiol.* **91,** 313–322.
MÜLLER, D. G. and JAENICKE, L. (1973). *FEBS Lett.* **30,** 137–139.
MÜLLER, D. G., JAENICKE, L., DONIKE, M. and AKINTOBI, T. (1971). *Science* **171,** 815–817.
MÜLLER, F. (1911). *Jahrb. Wiss. Bot.* **49,** 421–521.
MÜLLER, H. and MÜLLER, D. G. (1974). *Biologie in Unserer Zeit* No. 4, 97–105.
NAITOH, Y. and ECKERT, R. (1969a), *Science* **164,** 963–965.
NAITOH, Y. and ECKERT, R. (1969b). *Science* **166,** 1633–1635.
NAITOH, Y. and KANEKO, H. (1972). *Science* **176,** 523–524.
NIEDERPRUEM, D. J., JERSILD, R. A. and LANE, P. L. (1971). *Arch. Mikrobiol.* **78,** 268–280.
NIEDERPRUEM, D. J. and WESSELS, J. G. H. (1969). *Bact. Rev.* **33,** 505–535.
NUTTING, W. H., RAPOPORT, H. and MACHLIS, L. (1968). *J. Am. Chem. Soc.* **90,** 6434–6438.
OLSON, L. W. and FULLER, M. S. (1971). *Arch. Mikrobiol.* **78,** 76–91.
PARK, D. and ROBINSON, P. M. (1966). *In* "Trends in Plant Morphogenesis" (E. G. Cutter, ed.), pp. 27–44. Longmans, London.
PETTUS, J. A. and MOORE, R. E. (1971). *J. Am. Chem. Soc.* **93,** 3087–3088.
PFEFFER, W. (1884). *Unt. Bot. Inst. Tübingen* **1,** 363–481.
PLATTNER, J. J., BHALERAO, U. T. and RAPOPORT, H. (1969). *J. Am. Chem. Soc.* **91,** 4933.
PLATTNER, J. J. and RAPOPORT, H. (1971). *J. Am. Chem. Soc.* **93,** 1758–1761.
PLEMPEL, M. (1960). *Planta* **55,** 254–258.
PLEMPEL, M. (1962). *Planta* **58,** 509–520.
PLEMPEL, M. (1963). *Planta* **59,** 492–508.
PLEMPEL, M. and DAVID, W. (1961). *Planta* **56,** 438–446.

POON, N. H., MARTIN, J. and DAY, A. W. (1974). *Can. J. Microbiol.* **20,** 187–191.
RAI, P. V. and STROBEL, G. A. (1966). *Phytopathology* **56,** 1365–1369.
RAPER, J. R. (1939). *Science* **89,** 321–322.
RAPER, J. R. (1940). *Mycologia* **32,** 710–727.
RAPER, J. R. (1952). *Bot. Rev.* **18,** 447–545.
RAPER, J. R. (1957). *In* "Biological Action of Growth Substances" (Porter, H. K., ed.), pp. 143–165. Eleventh Symposium of the Society for Experimental Biology. Cambridge University Press.
RAPER, J. R. (1966). "Genetics of Sexuality in Higher Fungi". Ronald Press, New York.
RAPER, J. R. (1967). *In* "Handbuch der Pflanzenphysiologie" (W. Ruhland, ed.), Vol. 18, pp. 214–234. Springer, Berlin.
RAPER, J. R. (1970). *In* "Chemical Ecology" (E. Sondheimer and J. B. Simeone, eds.), pp. 21–42. Academic Press, New York.
RAPER, J. R. (1971). *In* "Plant Physiology" (F. C. Steward, ed.), Vol. 6A, pp. 167–222. Academic Press, New York.
RAWITSCHER-KUNKEL, E. and MACHLIS, L. (1962). *Am. J. Bot.* **49,** 177–184.
REINHARDT, M. O. (1892). *Jahrb. Wiss. Bot.* **23,** 478–563.
ROBINSON, P. M. (1973a). *Trans. Brit. Mycol. Soc.* **61,** 303–313.
ROBINSON, P. M. (1973b). *New Phytol.* **72,** 1349–1356.
ROBINSON, P. M. (1973c). *Bot. Rev.* **39,** 367–384.
ROBINSON, P. M., PARK, D. and GRAHAM, T. A. (1968). *J. Exp. Bot.* **19,** 125–134.
ROYLE, D. J. and HICKMAN, C. J. (1964a). *Can. J. Microbiol.* **10,** 151–162.
ROYLE, D. J. and HICKMAN, C. J. (1964b). *Can. J. Microbiol.* **10,** 201–219.
SNYDER, S. H., YANAMURA, H. I., PERT, C. B., LOGAN, W. J. and BENNETT, J. P. (1973). *In* "New Concepts in Neurotransmitter Regulation" (A. J. Mandell, ed.), pp. 195–222. Plenum Press, New York.
STADLER, D. R. (1952). *J. Cell. Comp. Physiol.* **39,** 449–474.
STADLER, D. R. (1953). *Biol. Bull.* **104,** 100–108.
TROUTMAN, J. L. and WILLS, W. H. (1964). *Phytopathology* **54,** 225–228.
TSUBO, Y. (1957). *Bot. Mag., Tokyo* **69,** 1–6.
TSUBO, Y. (1961). *J. Protozool.* **8,** 114–121.
WARD, H. M. (1888). *Ann. Bot.* **2,** 319–382.
WERKMAN, B. A. and ENDE, H. VAN DEN (1973). *Arch. Mikrobiol.* **90,** 365–374.
WESSELS, J. G. H. and NIEDERPRUEM, D. J. (1967). *J. Bact.* **94,** 1594–1602.
WIESE, L. (1965). *J. Phycol.* **1,** 46–54.
WIESE, L. (1969). *In* "Fertilization" (C. B. Metz and A. Monroy, eds.), Vol. 2, pp. 135–188. Academic Press, New York.
WOLPERT, L. and GINGELL, D. (1968). *In* "Aspects of Cell Motility" (Miller, P. L., ed.), pp. 169–198. Twenty-second Symposium of the Society for Experimental Biology. Cambridge University Press.
YANAGISHIMA, N. (1969). *Planta* **87,** 110–118.
ZENTMEYER, G. A. (1961). *Science* **133,** 1595–1596.
ZENTMEYER, G. A. (1966). *Phytopathology* **56,** 907.
ZENTMEYER, G. A. (1970). *In* "Root Diseases and Soil-borne Pathogens" (T. A. Tousson, R. V. Bega and P. E. Nelson, eds.), pp. 109–111. Univ. Calif. Press, Berkeley.

ZICKLER, H. (1937). *Z. Indukt. Abstammungs-Vererbungsl.* **73,** 403–408.
ZICKLER, H. (1952). *Arch. Protistenk.* **98,** 1–70.
ZIEGLER, H. (1962a). *In* "Handbuch der Pflanzenphysiologie" (W. Ruhland, ed.), Vol. 17, Part 2, pp. 396–431. Springer, Berlin.
ZIEGLER, H. (1962b). *In* "Handbuch der Pflanzenphysiologie" (W. Ruhland, ed.), Vol. 17, Part 2, pp. 484–532. Springer, Berlin.

Chapter 6

Chemotaxis of Leucocytes

P. C. WILKINSON

Department of Bacteriology and Immunology, University of Glasgow, Scotland

I. Introduction

The leucocytes of mammals are a specialized class of cells with a major function in defence against infection or tissue injury. In the broadest view, the function of these cells may be seen as the maintenance of the

TABLE 1. The leucocytes of mammalian blood and tissue fluids

Cell-type	Function	Chemotactic response demonstrated
Myeloid cells (*Granulocytes or polymorphonuclear leucocytes*)		
Neutrophil leucocytes	Short lived cells with major role in defence in acute inflammations and infections	Yes
Eosinophil leucocytes	Intestinal parasitic infestations Immediate hypersensitivity Lymphocyte-mediated hypersensitivity	Yes
Basophil leucocytes	Present in small numbers only. Release histamine and other vasoactive substances	Yes (One report only on leukaemic forms, Kay and Austen, 1973)
Mononuclear phagocytes (*Reticuloendothelial cells*)		
Blood monocytes	Precursor cell of tissue macrophage	Yes
Tissue macrophages (peritoneal cavity pulmonary alveoli etc.)	Long lived phagocytic cells especially seen in chronic inflammations, cell-mediated immunity, reactions to foreign bodies (tattoos) etc. Important role in carrying antigen to lymphocytes	Yes
Lymphocytes	Specific immunity. Antibody formation (B cells) Cell-mediated immunity (T cells)	No (but see footnote to p. 207)

integrity of the self. Cells which distinguish 'self' from 'not self' or from 'altered self', or which recognize damaged cells or molecules, require a sensory mechanism capable of making these distinctions at the molecular level. One class of leucocytes, the lymphocytes, which mediate specific acquired immunity, achieve this by their capacity to synthesize immunoglobulin antibody possessing structurally defined combining sites such that each antibody has a high stereospecific affinity for its complementary antigen which allows it to bond non-covalently with that antigen but not with unrelated antigens. Since different clones of lymphocytes produce antibody of different specificities, the body possesses an armoury of cells capable of responding specifically to a myriad of antigens.

The lymphocytes are not, however, the concern of this chapter, since, although lymphocytes are actively motile cells, tactic responses have not been shown in them, and the nature of their apparent directional migration has yet to be demonstrated.* Here we are concerned with the phagocytic leucocytes, namely those of the myeloid series, chiefly the neutrophil and eosinophil leucocytes, and those of the mononuclear phagocyte series, chiefly the blood monocytes and the motile macrophages found in tissue exudates (Table 1). These cells do migrate in gradients of soluble, diffusible substances. They have an important function as phagocytic cells and are characterized by the possession of a full complement of lysosomal hydrolases which are used for the destruction and digestion of endocytosed material. They do not possess antibody-like receptors, except when 'cytophilic' antibody produced by lymphocytes becomes bound to their surfaces. Therefore one of the central problems to be discussed in this review is how these cells, without the help of such receptors, distinguish substances which normally occur in the *milieu intérieur* and which they ignore, from those which occur only on damage or infection of tissue and which require to be removed.

It is apparent from the above observations that locomotion of leucocytes fulfils a very different function from that of the cells discussed in other chapters of this book. Thus the molecular recognition system of a phagocyte would not be expected to resemble those required for nutrition, reproduction or organization of tissues. In fact nearly all of the leucocyte chemotactic factors which have been described to date have been macromolecules, usually proteins or peptides, in contrast to the small molecules which are chemotactic in many other systems.

The word chemotaxis is used in this review in the sense of McCutcheon's definition (1946) as 'A reaction by which the direction of locomotion of cells or organisms is determined by substances in their environment' although, as will become apparent, this definition does not

* B-lymphoblasts show chemotactic responses (Russell, R. J., Wilkinson, P. C., Sless, F. and Parrott, D. M. V. (1975), *Nature (London)*, in press).

describe the locomotor response of leucocytes to chemical substances fully. The chemotactic responses to be discussed are positive—the cells migrate in a gradient towards the source of the attractant. Negative chemotaxis of leucocytes (away from the source of the gradient) is not known.

II. Methods for the study of leucocyte chemotaxis

Methods for studying leucocyte chemotaxis are of two types. Either the locomotor behaviour of individual leucocytes is observed in the presence or absence of attractant substances or in the presence or absence of a chemotactic gradient, or the effect of potential attractants on the behaviour of a population of cells is observed. It is a good deal easier to determine the chemotactic activity of unknown substances or to compare the activity of different substances using the second type of method—for example, Boyden's micropore filter method—than using the first.

A. The micropore filter method

Boyden (1962) introduced a convenient and reproducible method for assaying leucocyte locomotion by observations of the behaviour of a cell population. He used a two-compartment chamber in which the cells under test were separated from the chemotactic substance by a micropore filter. The cells were placed in one compartment above the filter and the chemotactic substance was placed in another below. The filter pore size was selected so that cells could not fall passively through the pores, but they could squeeze their way through them by actively migrating. Thus the pore diameter was usually somewhat less than the diameter of a cell at rest. At the commencement of the experiment, the concentration of the chemotactic factor was maximal below the filter, but zero within and above it. Molecules of chemotactic factor diffused through the filter from below to reach the cells on the upper surface, which responded by migrating into the filter. Thus a concentration gradient of chemotactic factor was set up within a few seconds of beginning the experiment. Zigmond (1974) has calculated the time required to set up a gradient of a molecule the size of albumin as 45 seconds. Such a gradient of a peptide or protein across a micropore filter of about 150 microns thick (the usual thickness of most commercially available filters) would gradually decay but would retain an appreciable steepness during the few hours or less necessary to measure leucocyte migration. Boyden's method, in which the number of leucocytes per unit area of the lower surface of the membrane after a certain lapse of time is measured, has proved ideal for the identification of chemotactic factors, for measuring their activity and for determining their cell-specificity. During the past decade it has been the method almost universally used by workers in the field. It measures the behaviour of a population of cells and is unsuited to the study of locomotion of single cells. For a single cell of the population the method measures only one of the velocity components, and gives only the average velocity.

It has been widely assumed that because cells migrated into a filter in response to factors diffusing from below, they were showing directional migration in response to a concentration gradient. In fact this is not necessarily so, because the cells may equally well be responding to the local absolute concentration of the attractant substance and not detecting concentration differences, i.e. they may be responding by chemokinesis rather than chemotaxis. It is in fact quite difficult to show that migration into micropore filters is chemotactic. This point is discussed below.

B. The slide and coverslip method

Because of these restrictions, there has been a revival of interest in methods which measure the behaviour of single cells. These methods have been used for many years (Comandon, 1917, 1919; McCutcheon, 1946; Harris, 1953a and b), but fell into desuetude because Boyden's method proved more convenient for answering the questions about the identity of leucocyte chemotactic factors and their role in inflammation which have occupied most workers in recent years. The most widely used of these methods is one in which a suspension of cells is allowed to spread between a slide and a coverslip and the subsequent migratory behaviour of the cells relative to a chemotactic source is studied, usually by time-lapse cinematography. One limitation of this method is that the chemotactic factor has to diffuse from a solid source. Clumps of bacteria have often been used as a source, since chemotactic factors are released by bacteria, but the rate of their release and the magnitude of the concentration gradient are unpredictable and difficult to measure. No satisfactory method has yet been developed for setting up reproducible concentration gradients with a slide and coverslip without a solid source. This greatly limits the uses of the method. Grimes and Barnes (1973) have recently made an adaptation in which a streak of a soluble chemotactic factor was allowed to dry on a slide and then to redissolve when a drop of cell suspension was added to it. However, it cannot be assumed either that the redissolution so achieved will give rise to a reproducible gradient from test to test, or that the structure of chemotactic substances is unaffected by such treatment.

III. The locomotion of leucocytes

A. Direct observations of migrating cells

McCutcheon and his colleagues (1946), using the slide and coverslip technique, attempted to quantitate the directional locomotion of leucocytes relative to a test object. They determined what they called a 'chemotactic ratio' of the net distance travelled by a leucocyte towards or away from a test object against the total distance travelled by the leucocyte. If, for instance, a leucocyte was 100 μm from the object, but its total path towards it (with changes of direction) was 200 μm, then the chemotactic ratio would be 100/200 i.e. +0·5—the plus sign indicating that chemotaxis was positive. Whether or not the leucocyte reached the object, the same ratio, net distance travelled/total distance travelled, was used. The limiting values for the ratio were +1·0 (movement in a straight line to the object) and −1·0 (movement in a straight line away from the object). With this method McCutcheon established that in the absence of a chemotactic substance leucocytes migrated randomly. If a chemotactic substance, e.g. a clump of bacteria, was present, almost all of the cells moved towards it, not in a straight line, but in a path which showed little deviation and became straighter as the cells approached the bacteria. When the cells reached the bacteria, they stopped moving. In this system leucocytes detected chemotactic substances at about 1 mm distance.

Robineaux and Frédéric observed the morphological changes in migrating leucocytes (Frédéric and Robineaux, 1951; Robineaux, 1954; Robineaux and Frédéric, 1955; Robineaux, 1964). They emphasized the appearance of a flattened veil or hyaloplasmic membrane which formed at the anterior end of the cell prior to locomotion, and which extended over the substratum on which the cell was migrating. The cell contents then flowed into it, leaving a long tail terminating in a number of adhesive filaments which were eventually broken by the force of the forward movement of the cell.

Ramsey (1972a and b), studying human neutrophil leucocytes, also emphasized the importance of formation of a hyaloplasmic membrane or 'lamellipodium', a flattened anterior extension of the cell, as a preliminary to locomotion. Lamellipodium formation was followed by cytoplasmic streaming into the lamellipodium. When a chemotactic gradient was present, the cell moved directionally in it. Ramsey considered that the event controlled by the gradient was not lamellipodium formation which he stated occurred randomly at any point on the cell surface irrespective of the direction of the gradient, but cytoplasmic streaming, since the cell contents flowed preferentially into lamellipodia on the side of the highest concentration of the attractant.

Both Ramsey and McCutcheon concluded that the presence of a chemotactic substance did not influence the speed of migration of leucocytes, but changed the nature of the migration from a non-directional to a directional form. However, other observations are contradictory to this. Keller and Sorkin (1966) showed, by photomicrographic tracing, that leucocytes in Gey's solution plus serum albumin, a non-chemotactic medium, moved very little during a 15 minute period. If heat-aggregated gamma-globulin plus 5 per cent serum, i.e. a chemotactic system, were present in the medium in an even concentration with no gradient, the cells showed greatly increased non-directional migration. The medium which McCutcheon had used had contained serum or plasma. Normal serum may contain substances which enhance leucocyte locomotion (Borel and Sorkin, 1969; Wilkinson *et al.*, 1969). This may account for McCutcheon's finding of considerable migration in experiments in which no source of a chemotactic gradient was added.

The most recent studies of the migration of individual leucocytes have been those of Zigmond and Hirsch (1973). These authors used human or horse neutrophil leucocytes and a slide and coverslip technique in which a streak of aggregated gamma-globulin was drawn across a coverslip and then allowed to dry. The coverslip was inverted over a slide containing a loopful of the cell suspension. The authors showed that the cells which came initially into contact with the gamma-globulin released a soluble chemotactic factor which diffused out to form a gradient and which could thus attract other cells. Release of this factor from leucocytes on the gamma-globulin streak formed a convenient system for observation of the behaviour of the cells at some distance from the streak.

In this system, the first event to occur in cells which were to migrate towards the centre of the gradient was that they became orientated. A lamellipodium formed at the front of the cell (nearest the gradient source) and a knoblike tail at the rear (Fig. 1). The orientation of cells was used as an indicator of directional migration, since this orientation was unmistakable and because time-lapse studies showed that cells moving towards a source were always orientated in this way during chemotaxis. Cells nearest the streak became orientated first and began to move towards the streak. As time passed, cells further away were also stimulated to move. This process could be observed as far away as, but not further than, 8 mm from the streak. After two hours over 90 per cent of the cells 2 mm from the streak were moving and 70 per cent of these were orientated towards the streak. In a well-developed response, individual cells showed chemotactic ratios—as defined by McCutcheon (1946) —of $+ 0{\cdot}61$ to $+0{\cdot}97$ with a mean of 0·85. The cells thus showed a positive directional response. They showed frequent changes of direction, but the direction of their turns was not random. There was a high probability that any cell which was moving at an angle more than 30° from the direct path would make its next turn in the direction of the source. Cells which reached the streak became immobile and spread out with their nucleus and granules packed into a central area.

Although the authors did not measure rates of locomotion in the presence and absence of a gradient, their results indicated firstly that

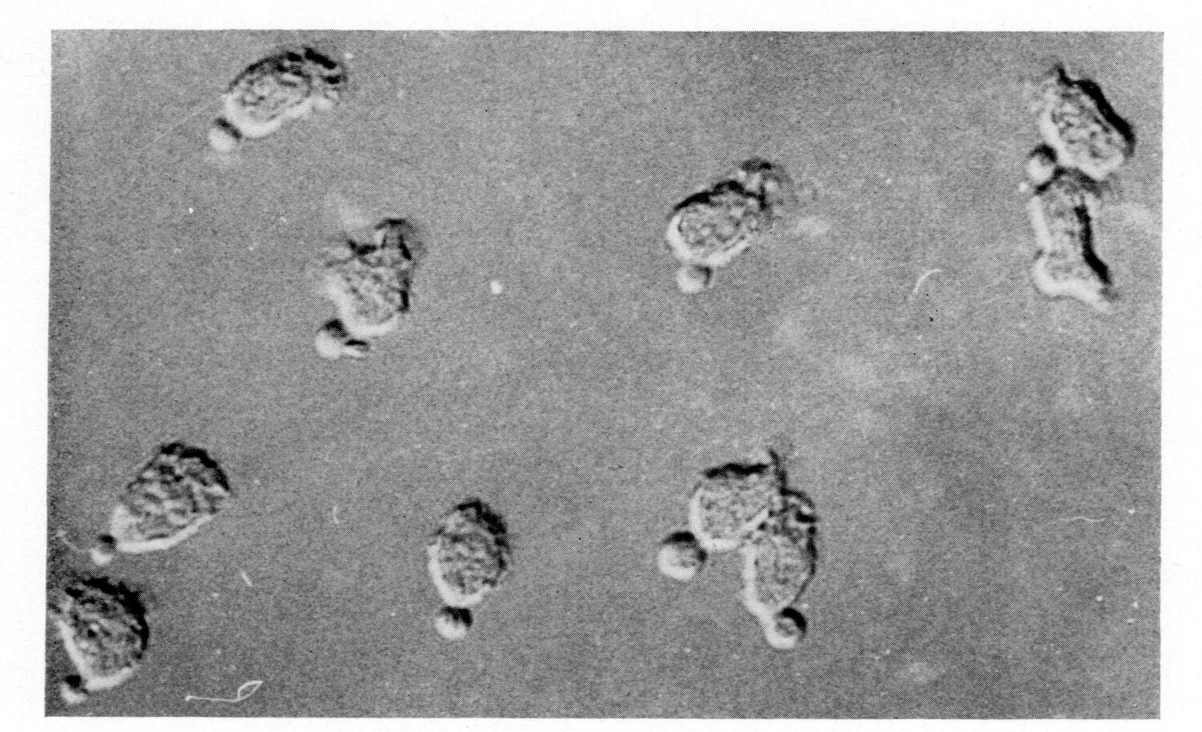

Fig. 1. Phase contrast photomicrograph of horse blood neutrophils migrating towards a chemotactic stimulus just beyond the top edge of the picture. Note the orientation of cells with a flattened anterior surface and a knob-like tail at the rear. Note also that most of the cells are orientated towards the upper margin of the field. Approx. ×400. From Zigmond and Hirsch (1973) by courtesy of Dr Sally Zigmond.

more cells migrated in the presence of a chemotactic factor than in its absence and secondly that cells in the presence of chemotactic factor, but not in a gradient sufficient to induce directional migration, did nevertheless show enhanced locomotion. Their results would therefore tend to support the observation of Keller and Sorkin (1966) quoted above that chemotactic factors increase the rate of locomotion of cells. They appear to do this even in the absence of a gradient, and, when there is a gradient, they influence the direction of movement as well.

A cell responding to a concentration gradient must be able to sense differences in concentration of the attractant substance. Two possible mechanisms have been suggested. Firstly the cell might be able to detect concentration differences across its length. This mechanism is called spatial sensing of a gradient. To do this, the cell requires at least two recognition sites, or a single site which can move over the cell surface. Secondly the cell might be able to read the concentration at a given point, move to a second point to take a second reading and, based on the difference between the two readings, make its next turn up the gradient (in positive chemotaxis) or down it (in negative chemotaxis). This mechanism, which would require some sort of memory, has been termed temporal sensing. Observation of the detailed turning behaviour of cells in gradients should make it possible to distinguish between these two sensing mechanisms. These observations have been made by Zigmond (1974b). The best evidence was obtained by observing the initial orientation of cells placed in a chemotactic gradient, before the cells began to move. Leucocytes did orientate themselves correctly before making any locomotor response. Eighteen out of 20 cells placed in a gradient orientated themselves towards, and made their first translocation into, the 180° sector towards the chemotactic stimulus. Such an orientation would occur by chance less than once in a thousand times. This would suggest that the stationary cell can detect concentration differences across its own length. Analysis of the frequency and direction of turn of leucocytes moving in gradients supports this conclusion. The length of cell path between turns did not alter as a function of the orientation of locomotion. However, the further a cell deviated from the direct path to the source the greater the magnitude of the next turn tended to be. Moreover, as mentioned above, it was usually towards the source.

Such findings favour, or are at least compatible with, a multiple site theory of chemotactic recognition. In a gradient, the direction of migration may be determined by the number of 'hits' by chemotactic factor on cell surface recognition sites. If the hits are distributed randomly over the whole surface, the cell will respond randomly. If, on the other hand, hits are more frequent on sites at the front end of the cell than at its rear, the cell will continue to move forward. It is not possible yet to say whether a cell in a gradient compares the amount of stimulation received by its various sides or whether stimulation at one site is followed by a refractory period which inhibits lamellipodium formation on other portions of the cell.

B. Directed and non-directed locomotion in micropore filters

In order to assess critically the contribution made to migration of cells in Boyden chambers by chemotaxis and by non-directed migration (chemokinesis), it is necessary to find a way of studying the migration of a cell population quantitatively. The most widely used method has been to allow the cells to migrate through the filter until they reach the lower surface and then to count the number of cells per high power field on that surface, in other words, to sample the population which has travelled a predetermined distance, i.e. to the bottom of the filter. However, this method has been shown to give rise to rather large errors, particularly since cells may fall off the lower surface (Keller *et al.*, 1972). Zigmond and Hirsch (1973) have made a successful adaptation using a different sampling technique. The experiment was stopped before the cells had time to reach the lower surface and locomotion was assayed by measuring the distance which the leading front of cells (two cells or more in the same focal plane) had migrated in a given time. This was shown to be a valid measure since the position of the leading front of cells was representative of the distance travelled by the whole migrating population (Zigmond and Hirsch, 1973). There was a considerable improvement in reproducibility over the previously-used counting method.

This technique was used to evaluate the contribution made by directed and non-directed locomotion as shown in Table 2. The experiment shown there was done in the author's laboratory but is based on similar experiments of Zigmond and Hirsch (1973). Varying concentrations of an attractant (in this instance, casein) were placed above and below the filter of a Boyden chamber. The cells were placed above the filter. If Table 2a is examined, it will be seen that above the diagonal from upper left to lower right, the cells are migrating from a low concentration to a high concentration of casein, i.e. in a positive gradient. Below this diagonal, they are migrating from a high concentration to a lower one, i.e. in a negative gradient. Along the diagonal, the concentration is the same on both sides of the filter. There is no gradient but the absolute concentration of attractant increases as one moves down the diagonal.

The first point to note on looking down the diagonal is that neutrophil leucocyte migration increases as the concentration of the attractant is raised, reaches a maximum and then decreases as the concentration

TABLE 2. (a) Migration of human blood neutrophil leucocytes towards a chemotactic substance (casein) in varying concentration gradients and at varying absolute concentrations.

		Distance travelled (μm) from upper towards lower surface of filter in 65 minutes.* Concentration of casein below filter (mg/ml)				
		0	0·2	1·2	2·2	3·2
Concentration of casein above filter (mg/ml)	0	23				
	0·2		46	82	76	89
	1·2		43	72	87	94
	2·2		42	61	84	100
	3·2		40	49	56	64

* Mean of three filters; 3 μm pore size.

(b) Calculated migration expected in the same experiment on the basis of non-directed locomotion alone.

		Estimate of distance travelled (μm) Concentration of casein below filter (mg/ml)				
		0	0·2	1·2	2·2	3·2
Concentration of casein above filter (mg/ml)	0	23*				
	0·2		46*	59	59	63
	1·2		59	72*	78	74
	2·2		77	78	84*	74
	3·2		72	74	74	64*

* Observed rates of locomotion (Table 2a) for cells in the absence of a gradient. The remaining figures were calculated as described by Zigmond and Hirsch (1973) on the basis of these figures. Note that above the diagonal (upper L to lower R) the calculated figures are lower than those actually observed, and below the diagonal they are higher.

is raised still higher. This effect has been shown with several attractants (see also Fig. 2). The rate of leucocyte locomotion is therefore sensitive to the absolute concentration of an attractant in the absence of a gradient. Gradient effects can be assessed roughly by comparing the figures above the diagonal in Table 2a (positive gradient) with analogous figures below the diagonal (negative gradient), or, perhaps more easily, by studying the figures along the diagonals from *lower* left to *upper* right. Along any of these diagonals, the mean concentration of attractant does not change, but there is a change from a negative gradient at lower left to a positive gradient at upper right. On moving along the diagonals, it can be seen that cell migration is enhanced as the gradient changes from negative to positive. However, although the mean concentration across the filter is uniform, the cells are not at all times

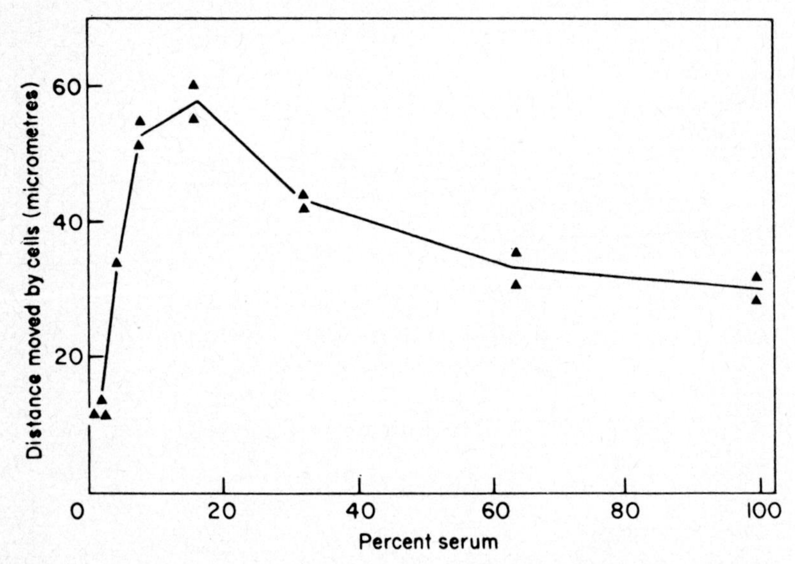

FIG. 2. Dose response curve of horse blood neutrophils migrating in a micropore filter in autologous horse serum. Ordinate: μm moved into filter by cells during 30 minutes incubation. Abscissa: Concentration of serum present both above and below the filter. From Zigmond (1974) by courtesy of Dr. Sally Zigmond.

exposed to this mean. Therefore a more exact method for measuring the effect of the gradient is needed. Zigmond and Hirsch (1973) and Knight (1974) have presented a method for calculating the expected migration of leucocytes into filters on the basis of a response to absolute concentration alone using the kind of experimental model shown in Table 2a. These calculations can be used to show how directional migration causes a deviation from such an expected result (shown in Table 2b). They provide substantial evidence that cells moving through micropore filters do indeed detect gradients. Leucocytes migrating in filters are therefore influenced both by the absolute concentration of an attractant and by the concentration gradient. However, a chemotactic effect can only be observed in filters when the absolute concentration of the chemotactic factor is optimal, i.e. on the rising slope and summit of the curve in Fig. 2. On the descending slope, directional effects are not seen. The migration of leucocytes in filters is therefore complex, more complex than the simple directional response suggested by McCutcheon's definition (1946).

One feature of note in Table 2 is that there is evidence of trapping of cells in a negative gradient. These cells move less than cells exposed to the same absolute concentration of attractant but without a gradient.

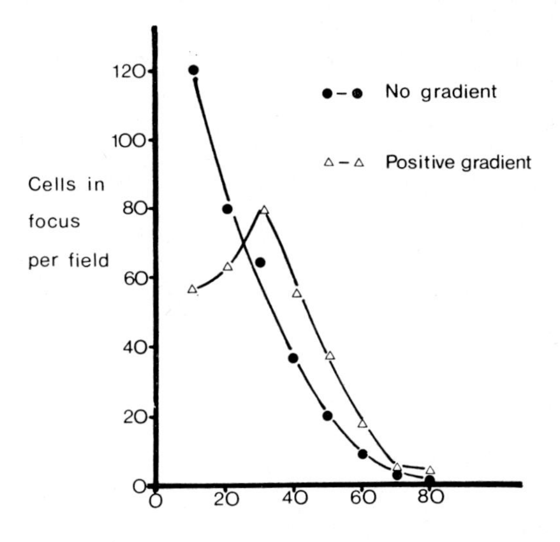

FIG. 3. Effect of a gradient on the distribution of cells migrating into a filter. ●—● No gradient. Alkali-denatured human serum albumin (HSA) 500 μg/ml above filter and 500 μg/ml below filter. △—△ Positive gradient. Alkali-denatured HSA 50 μg/ml above filter and 500 μg/ml below filter.

This may reflect the same phenomenon as the immobilization of cells seen at the centre of the gradient in the slide and coverslip studies described earlier.

In the inflammatory process, cells must migrate out of the blood vessels and through the tissue spaces towards the source of an attractant. It is probable that a migration which is partly directed by responsiveness to absolute concentration, partly by sensing a gradient, is more efficient *in vivo* than a migration which is purely chemotactic. The tissues of the body are not static, especially in inflamed sites, and it is probable that a gradient over any distance will become disturbed by movement of tissues. When the cells first leave the blood vessel and are still at some distance from the source of attractant, it is possible that their migration is largely random and its rate is determined by the local absolute concentration of the attractant. However, those cells which approach nearer to the source will come more and more under the influence of the gradient, which will be steepest and least disturbed close to the source. These cells will reach the source by chemotaxis, and having reached it they will be trapped. Thus it would seem that the nature of leucocyte migration as observed *in vitro* is well-adapted to its purpose *in vivo*. Observations *in vivo* are still technically too difficult to demonstrate this directly. In fact, one of the major concerns of workers in the leucocyte chemotaxis field is to show that the phenomenon does occur *in vivo* at all. There is some evidence that it does; see for instance Buckley (1963) who observed leucocytes migrating directionally through the tissues towards a microinjury, but such experiments are by no means easy to set up.

A different method for assessing the response of leucocytes to gradients in micropore filters is to study the final distribution of the cells through the filter. In the absence of a concentration gradient, the cells attain a normal distribution, i.e. they obey the equation for linear diffusion. Under such conditions, the peak of cell concentration will remain at the origin. In the presence of a positive gradient, the cell distribution is distorted from the normal, indicating a directional component in the migration. Zigmond and Hirsch (1973) showed that a gradient could influence the distribution of cells so that a peak of cells could be seen moving *en masse* through the filter. Fig. 3 shows the effect of a gradient on the distribution of cells in a filter.

IV. Cell-specificity in leucocyte chemotaxis

As mentioned earlier, chemotactic responses have been demonstrated in several different types of leucocyte. These include the neutrophil and eosinophil leucocytes and different types of mononuclear phagocyte including peritoneal and pulmonary alveolar macrophages and blood monocytes. In inflammatory responses of different types in the living animal, one or other of these cell types may predominate. The short-lived neutrophil leucocyte is the typical cell of acute inflammation, whereas macrophages, whose life-span is much longer, predominate in chronic granulomata, e.g. in tuberculosis. Eosinophils are present in greatest numbers in intestinal parasitic infestations and in the lesions of immediate hypersensitivity. Although there may be a number of factors which determine the cellular composition of a lesion, for instance the life-spans or mitotic rates of cells of various types, it is also probable that cell-specific chemotactic attraction controls the migration of the different cell-types into sites of inflammation. This is supported by studies *in vitro*. Certain chemotactic factors attract all three cell-types, neutrophils, eosinophils and macrophages. An example of such a factor is the complement peptide C5a (Snyderman *et al.*, 1969, 1971; Kay, 1970). Other factors show cell-specificity. For instance, a factor (Kay *et al.*, 1971) named ECF-A (eosinophil chemotactic factor of anaphylaxis), which is released from mast cells or basophils (Parish, 1972) when cell-bound IgE antibody or other homocytotropic antibodies react with antigen, attracts eosinophils preferentially, but may also attract neutrophils when tested in the absence of eosinophils (Kay *et al.*, 1973). Factors produced by granuloma-forming bacteria, e.g. mycobacteria (Symon *et al.*, 1972) and anaerobic corynebacteria (Wilkinson *et al.*, 1973a and b) attract macrophages preferentially. How is cell-specificity controlled? It might be postulated that there is a series of receptors on leucocytes, some of which are common to all cells and others of which are confined to certain cell types. The evidence to be presented below on the nature of recognition in leucocyte chemotaxis does not support this idea, and there are other objections to it, particularly the observation that a chemotactic factor can attract one type of cell under one set of conditions and another type of cell under another. Recent studies by Wissler *et al.* (1972) of a chemotactic system probably related to or identical to C5a have shown that in this system chemotactic activity is controlled by two peptides named CAT (classical anaphylatoxin) and CCT (co-cytotaxin). Neither of these peptides has activity on its own. However if the two peptides are both present, chemotactic activity is seen. Cell-specificity is controlled by the absolute concentrations and molar ratios of the two peptides. For example, at 5×10^{-5}M CAT + 1×10^{-7}M CCT, only neutrophils were attracted. At 1×10^{-9}M CAT + 1×10^{-5}M CCT, only eosinophils were attracted. At 5×10^{-6}M CAT + 5×10^{-7}M CCT, both cell types were attracted. This was a heterologous system (peptide from pig serum and cells from rabbit or guinea pig) and the results require to be confirmed in a homologous system. Nevertheless, it is difficult to account for such observations on the basis of cell-specific receptors. The control of cell-specificity is not understood and is still a major unknown area in the field of leucocyte chemotaxis.

V. Nature and sources of leucocyte chemotactic factors

A large number of chemotactic factors for leucocytes have now been described, nearly all as a result of using the Boyden technique. It is not the purpose of this review to discuss all of these in detail since there are other sources of information about them (Sorkin *et al.*, 1970; Wilkinson, 1974a). However a few generalizations may help the reader to understand the biological role of leucocyte chemotaxis. Chemotactic factors may be divided into two classes. The first of these, named *cytotaxins* by

Keller and Sorkin (1967a), are chemotactic factors proper which activate the cell directly, presumably by contact with a site or sites on its surface. These factors must be diffusible in order to set up a gradient. Many of them are exogenous, but many are endogenous and these include certain substances which are inactive in normal tissues, but may be converted to an active form by enzymes or other factors released in damaged tissues. The second group were named *cytotaxigens* by Keller and Sorkin (1967a). Cytotaxigens have no direct capacity to activate the cell. They are substances which can initiate changes in chemotactically inactive molecules such that chemotactic activity is conferred on those molecules. In many instances, they activate the endogenous cytotaxins mentioned above. Frequently, they need to be incubated with fresh serum or plasma at 37°C to manifest activity. This is because they activate humoral enzyme systems and liberate chemotactic products. Cytotaxigens do not require to be water soluble or diffusible, and many of them are particulate. Thus a chemotactic effect is exerted by many insoluble substances or by whole cells which could not be recognized and removed without the generation of soluble chemotactic factors. An example of a powerful cytotaxigen is the endotoxin (lipopolysaccharide) of the enterobacteria. This has no direct chemotactic activity, but if incubated in normal serum at levels as low as 0·05 μg per ml, a strong chemotactic activity is generated (Keller and Sorkin, 1967a; Snyderman *et al.*, 1968).

As would be expected many powerful exogenous cytotaxins are produced by pathogenic bacteria, although very little chemical characterization of these factors has been undertaken. They are reviewed elsewhere (Wilkinson, 1974a). The *endogenous* cytotaxins are of especial interest because they are activated whenever tissues are damaged. On any but the most minor injury to tissue, a number of proteolytic enzyme cascades are activated. The traumatic events convert a proenzyme to an active form which interacts with a protein substrate in such a way that this is itself converted to an enzymatically active form, and so on, so that by a series of enzymatic activations an initial event affecting only a few molecules may become greatly amplified. The cascades of importance for leucocyte chemotaxis are the complement system, the kinin system and the coagulation system. Chemotactic factors have been identified from all three, e.g. kallikrein from the kinin system (Kaplan *et al.*, 1972), fibrinopeptides and other factors from the coagulation system (Kay *et al.*, 1973; Stecher *et al.*, 1971; Stecher and Sorkin, 1972), but much the best studied is serum complement. The complement cascade is of especial importance in immunological reactions since it is activated by immune complexes containing complement-fixing antibodies. The best characterized of the complement chemotactic factors is the peptide C5a (Snyderman *et al.*, 1969) which is derived from the fifth component of complement by proteolytic hydrolysis, is liberated during cascade activation, and may be related to, or part of, the binary peptide system (Wissler *et al.*, 1972) described above. Two other complement chemotactic factors, a peptide C3a (Hill and Ward, 1969) and a macromolecular complex $C\overline{567}$ (Ward *et al.*, 1966) have also been described but their activity is less clearly established than that of C5a. These factors may be generated not only by cascade action but also by the direct action of proteolytic enzymes such as trypsin or plasmin on the precursor protein molecules. The active peptide C5a, for instance, may be released by proteolytic splitting of its parent molecule, C5 (Ward and Newman, 1969). This action is important because proteolytic enzymes are released from damaged or dead cells or from living leucocytes by physiological degranulation. Consequently, leucocytes are usually present in large numbers in areas of tissue necrosis, attracted presumably by peptides split from complement or other proteins by proteolytic enzymes released from cells (Borel *et al.*, 1969; Jensen *et al.*, 1969; Hill and Ward, 1969; Ward and Hill, 1970). Cells infected with viruses may release chemotactic factors (Brier *et al.*, 1970; Ward *et al.*, 1972) as may primed lymphocytes incubated with specific antigen (Ward *et al.*, 1969; Altman and Kirchner, 1972). Cell-derived soluble chemotactic factors are obviously important in inflammation and it is possible that cells also release inhibitors which regulate the entry of leucocytes into lesions. Migration is also regulated by the cytotaxin dose (Fig. 2).

Of the chemotactic factors discussed above those which have been characterized are proteins or peptides. There are few reports of chemotactic activity of non-peptide molecules. However we have recently solated a phospholipid from *Corynebacterium parvum* and allied species (a group which have immunopotentating effects when injected into experimental animals or man) which has a macrophage-specific chemotactic activity (Russell *et al.*, 1975). Cyclic AMP was reported to be inconsistently chemotactic for neutrophil leucocytes by Leahy *et al.* (1970). Although this suggestion has received some support, other laboratories including our own have not been able to repeat it. Tse *et al.* (1972) examined a whole series of purine and pyrimidine bases and concluded that cyclic AMP was not chemotactic but that intracellular cyclic AMP levels were important since substances which raised these levels in leucocytes depressed chemotactic responsiveness and *vice versa.* Conversely, a rise in intracellular cyclic GMP levels in leucocytes is associated with an enhancement of locomotor activity (Estensen *et al.*, 1973). Prostaglandin E1 is another non-peptide which has been reported to be chemotactic for leucocytes (Kaley and Weiner, 1971). Work in other laboratories has suggested that prostaglandins play a regulatory rather than a direct role by influencing the chemotactic response to other factors (Snyderman, 1973).

VI. Molecular recognition in leucocyte chemotaxis

The question of how leucocytes recognize and migrate towards many different and unrelated chemotactic factors is an intriguing one. These factors are so numerous that, although they are mostly peptides or proteins, any possibility of recognition of a detailed structural feature common to all of them is remote. Since it is unlikely that leucocytes possess specific recognition sites for all of these factors, it is necessary to look for common features about them which are not dependent on recognition of details of primary or folded structure. This search has been a primary interest of our laboratory recently. One could seek for clues in the structure of known and previously described chemotactic factors. Unfortunately none of these have yet been studied in enough detail for the necessary information to be available from which generalizations could be made. An alternative and more practical approach is to take non-chemotactic proteins and to attempt, by various physical and chemical manipulations, to render them chemotactic. This has been our own approach.

A. Studies of denatured proteins

Our first objective was to determine the effect of denaturation on the chemotactic activity of proteins as measured in the Boyden chamber. In the earliest experiments (Wilkinson and McKay, 1971) we used human serum albumin (HSA). This is a useful protein for such experiments; it is available in high purity and is present in large amounts in the natural environment of the human neutrophils used as test cells. Thus the interpretation of the experiments could not be confused by the possibility that the test substance might be antigenic or toxic. It is a single-chain protein, so that no change in quaternary structure could occur during denaturation.

HSA was denatured by several methods including change in pH or temperature or addition of chemical agents such as urea or guanidine hydrochloride. In general the simple physical methods were most informative. HSA treated with chemical denaturants such as guanidine HCl tended to aggregate on return to physiological conditions and was therefore not diffusible. The return to physiological conditions was of course necessary before the proteins could be tested for chemotactic activity against neutrophil or other leucocytes. This return was accompanied by a renaturation which was often incomplete and was always unpredictable. The physical state of the protein after denaturation and return to physiological conditions was therefore studied by measuring increases in its specific viscosity and surface activity and changes in its absorption spectrum in the UV region which are best detected by comparison with native HSA by difference spectroscopy. Changes in all of these properties are indicators of denaturation.

Denaturation caused HSA to become chemotactic for human neutrophil leucocytes. There was a good correlation between chemotactic activity and all three of the above physical parameters of conformational change, viscosity, surface activity and difference spectroscopy (Wilkinson and McKay, 1971, 1974; Wilkinson, 1974b). Where no conformational change was detectable, chemotactic activity was not observed. Those protein solutions which showed the grossest changes from the native form were also the most chemotactic. Usually alkali-denatured HSA (pH 12) was more chemotactic than acid-denatured HSA (pH 2–4) which was more chemotactic than heat-denatured HSA (70°C for 20 minutes).

On denaturation of a protein, the primary change is an alteration from the native conformation. However, polymerization may frequently occur as a secondary phenomenon. It was, therefore, necessary to determine whether the primary conformational change or the polymerization was the signal for recognition by the cell. This was achieved by separating polymer fractions of denatured HSA from monomer fractions by gel filtration on Sephadex (Pharmacia, Uppsala, Sweden). Monomer fractions were as active as or more active than polymer fractions, and therefore the conformational change is the essential one for recognition.

Haemoglobin and myoglobin were next studied (Wilkinson, 1973). These are suitable model proteins because their complete detailed structure is known and because, since they are haem proteins, they have a complex spectrum in the visible region, so that minor conformational changes often can be monitored by observation of spectral changes.

On binding of ligands such as oxygen or carbon monoxide, haem proteins undergo slight conformational changes. These changes are not accompanied by any acquisition of chemotactic activity. However, a major conformational change can be achieved if the haem group is removed from the globin molecule of haemoglobin or myoglobin. Haem is removed by precipitation of globin with acid-acetone at −20°C (Rossi Fanelli *et al.*, 1958). Its removal is accompanied by a reduction of α-helix content in the globin from about 70 per cent to about 55 per cent (Javaherian and Beychok, 1968; Breslow *et al.*, 1965; Harrison and Blout, 1965). The change occurs because haem is non-covalently bound in a 'hydrophobic pocket' and when it is removed, the exposure of many non-polar groups makes the native globin conformation thermodynamically unstable. Haem-free globin from human haemoglobin (apohaemoglobin) or from horse heart myoglobin (apomyoglobin) were both appreciably chemotactic for human blood neutrophils and monocytes.

If apohaemoglobin or apomyoglobin are mixed with haemin (the ferric form of haem), the protein and the prosthetic group spontaneously reassociate and the protein reassumes its native conformation indistinguishable from that of the original protein (Breslow *et al.*, 1965; Harrison and Blout, 1965; Javaherian and Beychok, 1968). This provided an excellent test system to determine whether the recovery of native form

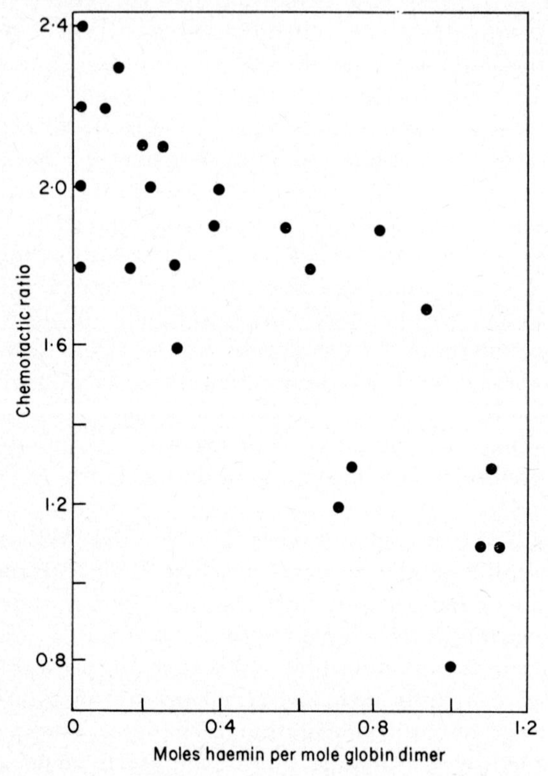

FIG. 4. The chemotactic activity for human neutrophils of apohaemoglobin combined with haemin at varying molar ratios. Chemotactic ratio = migration (μm in 75 mins.) towards test sample/migration towards native ferrihaemoglobin. Note that as the molar ratio of bound haemin is increased, the chemotactic activity of the protein decreases. From Wilkinson (1973).

was accompanied by loss of chemotactic activity. This should be true if leucocytes recognize deviations from the native conformation. The results are shown in Fig. 4. As haemin was added to apohaemoglobin (or apomyoglobin) at progressively increasing molar ratios of haemin to protein, there was a progressive loss of chemotactic activity of the protein. At a ratio of one mole of haemin per mole of globin dimer, chemotactic activity was completely lost. It is at this ratio that the native conformation is fully recovered as judged by optical measurements

(Javaherian and Beychok, 1968). These experiments therefore suggest that chemotactic activation of leucocytes depends on their recognition of changed conformation of a protein. If these changes can be reversed, the chemotactic activity of the protein is also reversed.

Neutrophil leucocytes and various types of macrophages were shown to respond to denatured proteins. However, we were unable to show such a response in eosinophils.

Enzymatic digestion might also be expected to give rise to products which were conformationally different from the undigested protein. Our own results using proteolytic enzymes have been somewhat inconsistent, but there are reports in the literature that enzymatic treatment of collagen (Chang and Houck, 1970; Houck and Chang, 1971) or of IgG (Yoshinaga *et al.*, 1972) yielded chemotactic products. In the case of IgG chemotactic activity was shown to correlate with a change in conformation of the molecule.

B. *Studies of proteins conjugated to synthetic side groups*

The next problem was to determine what it was about a denatured protein that a leucocyte recognized. One possibility was the recognition of the near random coil of a denatured protein in contrast to the unique folded form of the native protein. However this was ruled out by testing proteins or polypeptides which are random coils under physiological conditions. These include poly-L-glutamic acid and reduced-alkylated ribonuclease. These proved to be non-chemotactic. Linearity or randomness of structure are therefore not sufficient. The effects of charged groups on protein molecules are frequently investigated in cellular recognition studies. It is intrinsically unlikely that changes in electric charge would form a suitable signal for chemotactic recognition by leucocytes since proteins of widely varying charge are continually in contact with leucocytes, and charged groups are all externally positioned in the native conformation of haemoglobin and myoglobin so that denaturation cannot expose new charged groups. Polyanions such as poly-L-glutamic acid and polycations such as poly-L-lysine were not chemotactic. Furthermore, if the characteristics of the known chemotactic factors, such as complement peptides, casein and the denatured proteins cited above, are compared, there is no relation between chemotactic activity and isoelectric point. Neither net charge, nor individual charged groups are likely candidates to serve as chemotactic signals. Nevertheless, ionic groups on proteins probably do play an indirect role as discussed later.

The hypothesis which attracted us most was that the change recognized by leucocytes was the extroversion of non-polar groups on denatured protein molecules. These groups are packed into the interior of native hydrosoluble proteins and, indeed, their hydrophobic interactions provide most of the stabilizing energy which maintains the folded structure of such proteins in aqueous solutions (Kauzmann, 1959). On denaturation, these hydrophobic bonds are broken and non-polar groups become exposed to solvent. Such protein solutions show increased surface activity and would also be expected to show an enhanced affinity for hydrophobic sites on cell membranes.

We tested this hypothesis by conjugating a variety of small organic molecules to serum albumin under mild conditions in which denaturation was avoided or minimized. The presence of denaturation was checked by the methods detailed above. The side-groups conjugated to HSA were of varying character; negatively or positively charged, uncharged and polar or uncharged and non-polar. The degree of conjugation achieved was measured by absorption spectroscopy wherever possible; the infrared region was used for conjugates with aliphatic side groups and the visible or UV region for coloured aromatic conjugates such as azoproteins, fluorescein-conjugated proteins or picryl proteins.

TABLE 3. Chemotactic activity of some HSA preparations conjugated to different side groups.

Reagent conjugated* to HSA	Moles reagent per mole protein in reaction mixture	Chemotactic activity (human neutrophils)†.
Butyric anhydride	111	2·81
Propionic anhydride	134	2·49
Acetic anhydride	342	2·60
Succinic anhydride	175	2·08
Butyraldehyde	Very high	2·64
Formaldehyde	257	1·00
Tosyl chloride	92	2·29
Picryl chloride	56	1·66
Benzenediazonium Cl (benzylazo-HSA)	30	2·10
O-methyl isourea (guanidylation)	Very high	1·18
Iodoacetamide	90	1·44
Iodine + KI	55	1·29
Iodine (chloramine T method)		1·00

* See Wilkinson and McKay (1972) for conditions of conjugation and for fuller details of these and other conjugates.

† Migration (μm) in 75 minutes towards test protein ÷ migration in same time towards unconjugated human serum albumin.

In general, these experiments tended to favour the idea that superficially placed non-polar groups on a protein molecule were recognized and induced a locomotor response in leucocytes. A selection of results is shown in Table 3. These results are given in much fuller detail by Wilkinson and McKay (1972). The most chemotactic conjugates were those which contained bulky non-polar side-chains, e.g. tosyl-HSA or butyryl- or propionyl-HSA. If a negatively charged group was introduced into such a side chain, e.g. a succinyl group in place of a butyryl group, the activity was usually lower. Substitution of polar groups into aromatic rings usually reduced the chemotactic activity of the conjugate, e.g. picryl-HSA was less chemotactic than benzylazo-HSA (formulae shown in Fig. 5). Making HSA more basic by guanidylation

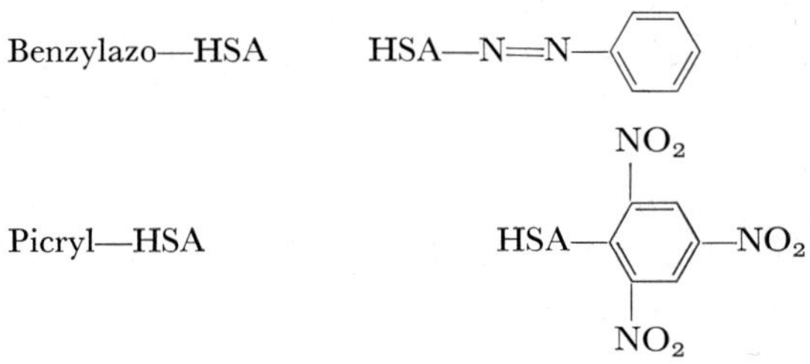

Benzylazo—HSA HSA—N=N—C_6H_5

Picryl—HSA HSA—$C_6H_2(NO_2)_3$

FIG. 5. Human serum albumin (HSA) is less chemotactic in conjugates in which the aromatic ring contains polar groups (e.g. picryl-HSA) than in those in which it does not (e.g. benzylazo-HSA).

did not make it chemotactic. Addition of small side groups, such as iodine or formaldehyde also made little difference to the chemotactic activity.

There was a positive correlation between the chemotactic activity of a conjugate and its side group: protein molar ratio (Wilkinson and McKay, 1972). The more non-polar groups there were attached to a protein molecule, the more chemotactic that protein was likely to be.

C. *Nature of the leucocyte recognition site for chemotaxis*

Since leucocytes responded to proteins conjugated to a variety of side groups whose molecular geometry had little in common, the problem posed earlier in relation to the multiplicity of chemotactic factors identified from biological fluids appears again in relation to conjugated proteins: how do leucocytes recognize a wide variety of structures which do not resemble one another in detail? There is, of course, the possibility that different conjugation procedures always caused an identical conformational change in the protein molecule. This seems unlikely and although it might apply to different HSA conjugates it would be unlikely to apply to other proteins. Conjugation to the groups discussed above was shown to enhance the chemotactic activity of other proteins such as immunoglobulins, ovalbumin or haemoglobin.

If there is a stereospecific receptor on the leucocyte with an affinity for the synthetic sidegroup in a chemotactic conjugate, then it ought to be possible to prevent the receptor from reacting with the chemotactic factor by adding to the chemotactic system the same synthetic substance in molar excess in free solution, but not by adding other substances whose detailed structure was different. This is essentially the type of experiment used by Landsteiner (1936) to study the inhibition of specific antigen-antibody interactions by free hapten. We studied the effect of the presence of tosylarginine methyl ester at 5×10^{-4}M on the chemotactic effect of tosyl-HSA at 1×10^{-5}M (Wilkinson and McKay, 1972). The tosyl groups are conjugated in a similar way in both protein and ester. In fact, at the concentrations quoted above, tosyl-arginine methyl ester has no effect on the chemotactic activity of tosyl-HSA or of HSA conjugated to other side groups. Thus it is unlikely that the leucocyte possesses a specific recognition site for tosyl groups. Similar conclusions were reached in experiments of the same type using other groups. Therefore we concluded that the recognition of chemotactic factors by leucocytes was not stereospecific. This conclusion is reasonable in physical terms if the presence of external non-polar groups on proteins is indeed the signal for recognition and if these groups form hydrophobic interactions with the cell surface. Hydrophobic interactions do not usually result in stoichiometric bonding, and, although they may be energetically stable, they show little directionality and are therefore not usually stereospecific.

The reaction of chemotactic proteins with the leucocyte membrane needs to be investigated further. At present, we see this interaction as one in which there is little stereospecificity. It is possible that the binding affinity of these proteins for the membrane may depend not on high affinity bonding at a single site, but on the combined effect of low affinity bonding at multiple sites.

D. The casein model

So far this discussion of recognition in leucocyte chemotaxis has been centred on proteins modified under experimental conditions. It is worth looking at previously-described chemotactic factors to see whether these conclusions can be applied to them as well. A very interesting and highly chemotactic protein is casein (Keller and Sorkin, 1967b) which attracts both neutrophils and macrophages. Casein is the major protein of milk. It is not a single protein, but a complex of which the major components are α_s-casein, β-casein and κ-casein. Since the chemotactic activity of casein components had not been determined, this was done in our laboratory (Wilkinson, 1972, 1974b). In summary these experiments showed that the α_s- and β-caseins were chemotactic but that κ-casein had little activity. The chemotactic activity of the α_s- and β-caseins was dependent on the casein being in low molecular weight form (monomer or small polymers). Casein tends to form micelles and it was found that the high molecular weight forms were not chemotactic. The chemotactic activity of casein was very dependent on Ca^{++} concentration. If this was raised, activity was lost. This is probably because raising Ca^{++} ion concentration favours the formation of micelles from low molecular weight soluble casein.

Since milk proteins are of major significance in nutrition and have economic importance, there is a considerable body of work on the chemistry of the caseins (see McKenzie, 1971). A model has recently been proposed for the structure of α_s- and β-caseins by Waugh *et al.* (1970). This model is of great interest for recognition in leucocyte chemotaxis. α_s-Casein and β-casein resemble one another closely but both are structurally unrelated to κ-casein. Both α_s- and β-caseins are acidic phosphoproteins. According to the model of Waugh *et al.* they contain an eccentrically positioned highly acidic peptide with a high proportion of serine phosphate monoesters and free carboxyl groups. This small peptide is linked covalently to a larger peptide which contains few charged groups but has a high content of aminoacids with non-polar side chains (40 to 50%). The authors suggest that there are too many non-polar side chains on this peptide for them all to be packed internally so some must be present at the surface of the molecule. Casein micelles are formed by intermolecular interaction of non-polar groups on adjacent molecules (Fig. 6). However micelle formation is limited by the presence of the acidic peptides. These repel each other and reduce the tendency towards intermolecular association. Any procedure which reduces their negative charge, e.g. dephosphorylation or binding to Ca^{++} ions, increases the tendency of casein to aggregate. We have shown that both of these procedures also reduce the chemotactic activity of casein. Casein in milk has low chemotactic activity. This is probably because of the high Ca^{++} concentration in milk. At the Ca^{++} levels found in serum, casein is strongly chemotactic.

The casein model illustrates some of the problems of chemotactic recognition. Proteins carrying exposed non-polar groups are unlikely to be thermodynamically stable in aqueous solutions and will either refold, if possible, or form aggregates. The tendency to aggregation or refolding could be countered if the protein molecule contained many similarly charged ionic groups. These would, by mutual coulombic repulsion, act in favour of an open conformation and inhibit intermolecular association. They would help to retain the non-polar sites in a solvent-exposed position accessible for recognition. This idea is favoured by some unpublished experiments of Dr Ian McKay and myself. The corn protein, zein, has a very high content of leucine and is insoluble in water. However, if zein is nitrated or sulphonated, it becomes soluble. Sulphonated zein contains many negatively charged sulphonyl groups and non-polar leucine side chains. It is chemotactic for neutrophils. However, nitrated zein, in which polar but uncharged nitro groups

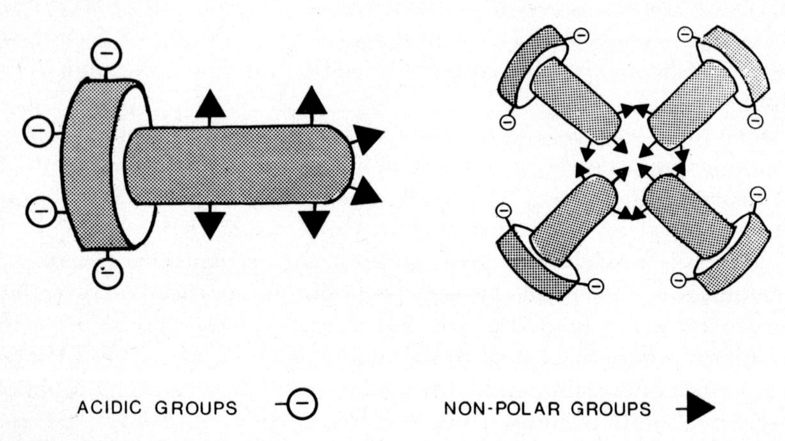

FIG. 6. Schematic model of the molecule of α_s- or β-casein. The monomer is shown on the left and its association into polymers on the right. After Waugh *et al.* (1970).

have been added, is not active. Thus a chemotactic protein of non-polar character, by the possession of mutually repulsive charged groups, may be stabilized in a form in which it can interact with cells.

E. Possible mechanisms of recognition

I have so far discussed recognition of chemoattractants by leucocytes in terms of the structure of the attractant molecule, and the conclusion of this discussion has been that there is no single structural feature which would account for recognition of these molecules. The recognition process, like many other membrane-related phenomena (Singer and Nicolson, 1972) may be better understood in thermodynamic than in structural terms. In considering protein denaturation, it may be stated that "the three-dimensional structure of a native protein in its normal physiological milieu is the one in which the Gibbs free energy of the whole system is the lowest" (Anfinsen, 1973). A denatured protein in aqueous solution will be in a higher free energy state than its native conformer. Such a protein might find its lowest free energy state at a surface rather than in the bulk of the solution and therefore be surface-active. An interaction with hydrophobic sites on cell membranes might also be thermodynamically favoured and there is, in fact, a correlation between surface activity and chemotactic activity of structurally altered proteins (Wilkinson, 1974b). Alternatively, transitions from the denatured to the native form will be thermodynamically favoured in solution under physiological conditions and Wissler and Sorkin (1973) have suggested that such transitions might be detected by leucocytes if they are slow and if the free energy change, $-\Delta G$, exceeds 10 kcal per mol. It is difficult to say more about this hypothesis while the detailed considerations on which it is based remain unpublished. It is based on calculations of ΔG for conformational transitions of proteins in aqueous solution. Recognition takes place however at the cell surface where the lowest free energy state of a protein is probably not that of the protein in aqueous solution.

My own prediction would be that proteins which are chemoattractant for leucocytes following modifications which enhance their hydrophobicity are capable of a more intimate interaction with the cell membrane bilayer than hydrophilic proteins, and that this intimate interaction is the event which triggers cell locomotion. These surface-active proteins are probably more stable thermodynamically at the cell surface than in aqueous solutions because their exposed hydrophobic sites can interact with similar sites within the bilayer. In Singer and Nicolson's fluid mosaic model of membrane structure (1972) the integral membrane proteins are embedded deep in the membrane by virtue of their content of non-polar regions which are most stable in the hydrophobic environment of the interior of the membrane. Amphipathic structure is a typical feature of integral membrane proteins (Singer and Nicolson, 1972) and the monomers of α_s- and β-caseins also have an amphipathic structure with one highly polar end and the other end non-polar. Bee venom melittin is also amphipathic and surface active but does not show chemotactic activity because it is very cytotoxic (Wilkinson, 1974b). It is strongly positively charged and therefore may bind avidly to negatively charged sites on the membrane and cause gross permeability changes. It has been suggested that chemotactic factors cause 'minimal irritation' to the cell membrane insufficient to cause cytotoxicity (Jenson and Esquenazi, 1975). The negatively-charged caseins may bind to the membrane less avidly than melittin and cause sufficient perturbation to increase its permeability to ions but not to disrupt it.

This hypothesis is supported by the finding that bee or snake venom phospholipases which increase membrane permeability can also, at non-toxic doses, enhance the response of leucocytes to chemotactic factors (Wilkinson, 1974b).

* This paper also suggests that antibody-coated leucocytes can show specific chemotactic responses to antigen.

In the introduction to this review, the necessity for complex organisms to distinguish self from not-self or from altered self was emphasized. In specific immunity a self: not-self distinction is made. The recognition system described here is not of the self: not-self type. Leucocytes from humans ignore native human serum albumin, but they also ignore native ovalbumin or haemocyanin, for example. A more accurate description of recognition in phagocytes would be based on their capacity to distinguish damaged from undamaged structure.

VII. Events in leucocytes which follow chemotactic recognition

How does sensory recognition of chemotactic factors by leucocytes activate a motor response in them? Answers to this question are beginning to emerge. The first studies of cellular events following contact with chemotactic factors were those of Becker and his colleagues who emphasized the importance of cell-bound serine esterases in chemotaxis (Ward and Becker, 1968; 1970a; Becker and Ward, 1969; Becker, 1971). On contact of the leucocyte with complement-derived or bacterial chemotactic factors, a proesterase (proesterase 1) is converted to an active esterase (esterase 1) which can be assayed biochemically. Becker's group believe this esterase activation to be an essential preliminary to locomotion. Their evidence is based on the specific inhibition of the activated form of this esterase by members of a series of phosphonate esters. If the esterase activity is blocked by such phosphonate esters, no locomotor response ensues in the cell. On the other hand Woodin and colleagues (Woodin and Wieneke, 1970; Woodin, 1971; Woodin and Harris, 1973), while agreeing that phosphonate esters can be shown to inhibit esterase activity in neutrophils, argue that this inhibition is irrelevant. They believe that organic phosphorus compounds act at the cell membrane by stabilizing the separation of hydrophobic surfaces there. Woodin suggests that the reversible separation of hydrophobic surfaces is essential for chemotaxis—an idea which might fit rather well with our own ideas of how chemotactic proteins interact with the cell membrane—and that this separation is made irreversible by the detergent action of phosphonate esters. The disagreement between the two groups has centred on how closely the specificity of different organic compounds as inhibitors of leucocyte locomotion can be related to the specificity of their action as esterase inhibitors or as detergents.

Woodin and Wieneke (1970) suggested that the site of action of phosphonates on the leucocyte is the potassium pump. A K^+- sensitive acyl phosphatase, whose action can be blocked with ouabain is believed to control cation movement across the leucocyte membrane, thus playing a similar role to that played by the Na^+K^+ sensitive ATPase present in other cells but absent from leucocytes. Phosphonates enhance the action of staphylococcal leucocidin which stimulates this enzyme, increases the cation permeability of the cell and causes orthophosphate to accumulate in the cytoplasm (Woodin and Wieneke, 1969). In the presence of Ca^{++} ions, this is followed by degranulation and release of β-glucuronidase and other proteins into the extracellular medium, deposition of calcium phosphate in cytoplasmic vesicles derived from the empty granules and the stimulation of a new orthophosphate-nucleotide reaction in the cell membrane. Leucocidin (in low doses) allows ion transport by reversibly separating two hydrophobic surfaces in the membrane. Woodin and Wieneke (1970) suggest that phosphonates inhibit chemotaxis and other functions by stabilizing this separation of hydrophobic surfaces.

Cation transfer across the cell membrane may be a necessary prerequisite for leucocyte locomotion as it is for transmission of nerve impulses or for muscle contraction. Leucocyte chemotaxis is little affected by changes in K^{+} concentration in the medium (Wilkinson, 1975), but is sensitive to changes in the concentration of divalent cations (Bryant *et al.*, 1966). Locomotion is diminished but not abolished when the medium is free of Ca^{++} and Mg^{++} (Becker and Showell, 1973; Wilkinson, 1975). Divalent cation ionophores like A23187 (Eli-Lilly) or X537A (Hoffmann-Laroche) are inhibitory to locomotion at high doses, probably because they increase the intracytoplasmic Ca^{++} concentration and induce a state of rigor in the leucocytes (Wilkinson, 1975). Probably leucocyte locomotion, like regulated muscle action, requires a flux of cations into the cytoplasm followed by their removal into intracytoplasmic vesicles analogous to the sarcoplasmic reticulum. At low doses, these ionophores, in the absence of extracellular Ca^{++} or Mg^{++}, restore the migration of neutrophils and monocytes to normal (Wilkinson, 1975). Since there is no Ca^{++} or Mg^{++} to transfer across the plasma membrane, this action of ionophores is presumably due to their ability to allow the leucocyte to make more efficient use of its own intracellular divalent cation stores. A calcium flux between the cytoplasm and the sarcoplasmic reticulum is necessary for contractile events in muscle. If leucocyte locomotion is mediated through contraction of microfilaments and controlled at the cell membrane it is possible that chemotactic factors interact within the phospholipid bilayer to initiate contraction by increasing the local flow of Ca^{++} across either the plasma membrane or the membrane of intracellular Ca^{++} storage vesicles. Enzymes controlling cation transfer may be sited on the inner side of the phospholipid bilayer; $Na^{+}K^{+}$ATPase in erythrocytes for example is sited on the cytoplasmic surface of the membrane (Marchesi and Palade, 1967). If a divalent cation-transfer enzyme were similarly sited and necessary for cell locomotion, this would explain why hydrophobic proteins, but not hydrophilic ones, can excite locomotion, since only a hydrophobic protein could penetrate deep enough into the membrane to contact a receptor on its inner surface. An explanation would also be required as to why low molecular weight surface agents, which could also penetrate the membrane, are not chemotactic, although they may be good inhibitors of chemotaxis, and why chemotactic factors are usually macromolecules.

The foregoing discussion strongly implies that leucocyte locomotion involves divalent cation-dependent contractile elements in the cell cytoplasm, namely microfilaments or microtubules. Microfilaments are demonstrable microscopically in macrophages (Bhisey and Freed, 1971; Allison *et al.*, 1971) and the microfilament proteins actin and myosin have both been isolated in pure form from neutrophils (Tatsumi *et al.*, 1973; Stossel and Pollard, 1973). Cytochalasin B, which disrupts microfilaments, inhibits locomotion of macrophages (Allison *et al.*, 1971) and neutrophils (Becker *et al.*, 1972; Zigmond and Hirsch, 1972). However, as Zigmond and Hirsch (1972) point out, cytochalasin B may act on functions other than microfilament contraction, so although it is very likely that the microfilament system is the machinery which drives the migrating leucocyte, independent evidence is still needed.

Microtubules are also present in leucocytes and in orientated and migrating macrophages they have been shown to form a cytoskeletal structure in the cytoplasm which maintains the polarized shape of the cell (Bhisey and Freed, 1971; Allison *et al.*, 1971). There is now a good deal of evidence that leucocytes depleted of microtubules still migrate well and show a chemokinetic response, but they cannot orientate themselves in a gradient and loose the ability to show chemotactic responses. Colchicine and vinblastine, which both specifically depolymerize microtubule proteins, both cause loss of the elongated, orientated form in migrating macrophages (Allison *et al.*, 1971; Bhisey and Freed, 1971). Bandmann *et al.* (1974) showed that the antitubulins demecolcine and podophyllic acid ethylhydrazine do not affect random migration of neutrophils but do diminish migration in a gradient of casein. I have used experiments of the pattern shown in Table 2 and shown (unpublished) that, after treatment with colchicine, the values for migration of both human blood neutrophils and blood monocytes are much nearer to those predicted for random migration than the values obtained for cells untreated with colchicine. Microtubules, therefore, appear not to be required for random locomotion, whether chemically stimulated or not, but are essential for orientated, tactic, locomotion.

The leucocyte which is activated to migrate by contact with a chemoattractant undergoes a metabolic burst which probably supplies the ATP necessary for contraction. The main energy source for leucocyte locomotion is anaerobic glycolysis. Inhibitors of anaerobic glycolysis such as iodoacetate or deoxyglucose (Carruthers, 1966; 1967) or iodoacetamide inhibit locomotion, but uncouplers of oxidative phosphorylation such as dinitrophenol or sodium arsenite (Carruthers, 1967; Bryant *et al.*, 1966) have only a feeble inhibitory effect. Since neutrophil leucocytes have very few mitochondria, this result is not unexpected. Leucocytes ingesting foreign particles show a metabolic burst characterized by increased glucose utilization, oxygen consumption and lactate accumulation, stimulation of the hexose monophosphate shunt and increased release of lysosomal hydrolases (Sbarra and Karnovsky, 1959; Iyer *et al.*, 1961; Cagan and Karnovsky, 1964; Karnovsky *et al.*, 1970). The respiratory rate of phagocytosing neutrophils or macrophages is higher than that of resting cells. Such a burst of activity in the glycolytic and hexose monophosphate shunt pathways accompanies not only phagocytosis but also chemotaxis (Goetzl and Austen, 1974). Such metabolically activated migrating leucocytes release much larger amounts of lysosomal enzymes such as acid phosphatase, β-glucuronidase and β-galactosidase than resting cells (Wilkinson *et al.*, 1973a; O'Neill, Cater and Wilkinson, unpublished). We also have evidence that activated neutrophils release increased quantities of proteolytic enzymes (Wilkinson and McKay, unpublished). The hydrolases are released from living leucocytes into the extracellular medium by exocytosis of cytoplasmic granules, and they can be assayed there after the leucocytes have been removed.

The pattern of recognition for enzyme release is similar to that for chemotaxis. If a number of denatured or conjugated proteins are tested for their capacity to enhance enzyme release, there is a close correlation between their ability to do this and their chemotactic activity. As suggested above, the same membrane site may activate both events. On the other hand, other tests of neutrophil metabolic function, e.g. the ability to reduce the dye nitroblue tetrazolium (NBT) which is dependent on stimulation of the hexose monophosphate shunt, do not show any correlation with chemotactic activity in our hands. Heat-denatured HSA for instance, enhances NBT reduction in neutrophils more strongly than does alkali-denatured HSA, whereas the reverse is true for chemotaxis. Heat-denatured HSA is polymerized but shows only slight conformational changes. Other workers (Nydegger *et al.*, 1973) have shown that the ability of immune precipitates to enhance NBT reduction is related to the size of the immune complex, so particle size rather than hydrophobicity may be the factor which regulates the NBT reducing system.

The ability of chemotactic factors to activate a metabolic burst in leucocytes may explain how these factors enhance non-directed locomotion, since the increase in locomotion may result directly from the increase in metabolic rate of the stimulated cells. A metabolic burst may be an essential prerequisite for enhanced locomotion and chemotaxis.

VIII. Chemotaxis and phagocytosis

Do relationships exist between the recognition mechanisms for chemotaxis and phagocytosis? In order to approach this question, it is necessary to summarize the evidence for recognition in phagocytosis. Macrophages, and probably also neutrophils, appear to employ at least three different recognition systems in ingesting particles (Rabinovitch, 1970; Cohn, 1972). These are: (1) The binding of IgG coated particles to a receptor on the leucocyte surface which recognizes the Fc fragment of IgG. This type of phagocytosis essentially involves a recognition of immune-complexes, the well-known phenomenon of opsonization: (2) There is a receptor for complement-coated particles (Lay and Nussenzweig, 1968). This receptor is destroyed by trypsinization and lost on culture of cells *in vitro* so it may not represent an integral membrane site but something adsorbed to the cell surface and rather easily detached. (3) Leucocytes have the capacity to recognize the surfaces of altered particles such as effete or aldehyde-treated erythrocytes, and to remove them by phagocytosis. This recognition is very efficient, as judged by the ability of hepatic Kupffer cells to clear particles from the circulation.

No studies of the nature of the molecular interactions involved in these recognition systems have been made, so no comparisons at the molecular level can be made with the chemotactic recognition system. It could be that they all depend on the same type of interaction. However, it is probable that the membrane site for binding immunoglobulin is different from that for phagocytosis of damaged red cells. If the macrophage Fc receptor is blocked by anti-macrophage IgG antibody in the absence of complement (Holland, Holland and Cohn, 1972), the cells no longer ingest IgG-coated particles. However they can still phagocytose polystyrene, aldehyde-treated erythrocytes or complement-coated erythrocytes. This suggests that different membrane sites are involved in mechanism (1) and mechanism (3).

It would seem that mechanism (3) above shows the closest parallel to chemotactic recognition. The processes involved in phagocytosis of damaged red cells or beads of non-polar materials such as polystyrene may be quite similar to the chemotactic recognition discussed above. This is given considerable support by some recent work of van Oss and Gillman (1972) and Thrasher *et al.* (1973). The idea that the interfacial tension between a cell and a particle presented to it for ingestion would determine whether the cell would phagocytose the particle was first explored by Mudd and Mudd (1924, 1933). They concluded that the degree of hydrophobicity of a particle was an important determinant of its phagocytosability. This idea was revived by van Oss and Gillman (1972) who used contact angle measurements to determine the interfacial tension of the surface of phagocytes and of particles. These measurements were achieved by estimating the contact angle of a water droplet on a monolayer of the test cells or particles. The higher the contact angle, the more hydrophobic is the cell or particle surface. They showed that, if the contact angles of neutrophil leucocytes and of different cultures of bacteria were measured, and this measure was correlated with the ability of the neutrophils to phagocytose the bacteria, only those bacteria which made a higher contact angle with a water drop than that made by neutrophil leucocytes were phagocytosed by the leucocytes. In other words, bacteria with water-repellent surfaces were more easily phagocytosed than those with hydrophilic surfaces. Capsulated *Staphylococcus aureus* is very hydrophilic and is phagocytosed very poorly. If the hydrophilic polysaccharide capsule is removed, the contact angle made by the bacteria increases and they become phagocytosable. The same phenomenon has been shown using macrophages (Thrasher *et al.*, 1973). Normal human erythrocytes have a lower contact angle than macrophages and are not phagocytosed by them. If the erythrocytes are coated with specific antibody, their contact angle is raised and phagocytosis is enhanced. The authors also presented evidence that primed lymphocytes release a factor which lowers the contact angle of macrophages and thus enhances the capacity of the macrophage for phagocytosis.

It is very likely that phenomena of this type are related at the molecular level to the activity of the chemotactic proteins discussed earlier. It should be possible to test this in phagocytic systems using particles to which side groups with well-defined physical properties have been covalently linked.

IX. Chemotaxis and pinocytosis

Recognition of pinocytosable molecules by macrophages is apparently completely different from recognition of chemotactic factors or from recognition in phagocytosis. Highly charged molecules such as polylysine or dextran sulphate are rapidly pinocytosed (Cohn, 1970, 1972). Pinocytosis of soluble proteins is also enhanced if the protein is polymerized or if antibody is added to form an immune complex. Neither charge nor aggregation is of much direct importance in chemotactic recognition. Macrophages show pinocytosis, but neutrophils apparently do not pinocytose material of less than 0·1 μm in diameter (discussion of paper by Cohn, 1970). The question should therefore be asked: why do neutrophils recognize and migrate towards soluble protein chemotactic factors if they lack the capacity for endocytosing them when they arrive? Of course, endocytosis of such proteins may take place by a mechanism other than that investigated by Cohn (1970). We have recently investigated the fate of ^{131}I-labelled denatured or conjugated proteins incubated with human blood neutrophils or monocytes (Wilkinson and McKay, unpublished). These experiments indicated that the cells could digest denatured serum albumin efficiently without endocytosis. This was achieved largely by extracellular release of proteases by the physiological mechanism discussed earlier. Unfolded proteins are especially susceptible to proteolytic attack because they present more substrate sites than a folded protein. There was no digestion of native serum albumin under similar conditions. Therefore the question may be answered, in part at least, by saying that endocytosis is not necessary for digestion of denatured proteins by neutrophils. On the other hand, exudate macrophages, in which pinocytosis is efficient, were not shown to release enzymes into the medium on contact with denatured proteins.

X. Conclusions

This review has concentrated on some current growing points in the study of leucocyte chemotaxis. The nature of chemotactic and non-chemotactic sensing of the presence of cytotaxins by leucocytes was discussed. Present evidence indicates that leucocytes sense, and are responsive to, both the absolute concentration of chemotactic factor—so that random migration is enhanced when there is no concentration gradient—and concentration differences—so that they migrate in a concentration gradient towards the gradient source. They probably sense gradients by detecting concentration differences across their own length. When leucocytes are placed in contact with chemotactic factors, whether or not in a gradient, they become metabolically activated as judged by several biochemical criteria. This metabolic burst accompanies, and is probably essential for, an acceleration in the cell's locomotion and may be seen as a necessary prerequisite for directional migration, phagocytosis and intracellular digestion. Thus these functions of the leucocyte are operationally linked.

The molecular basis for recognition of chemotactic factors by leucocytes was also discussed. A large number of chemotactic factors have been described and a unifying hypothesis is required to explain recognition of these many factors by cells which are not capable of immune responses of the specific adaptive type. From studies of a number of proteins modified by a variety of laboratory procedures, some constant features have emerged. Leucocytes are capable of detecting modifications of protein structure if these modifications are accompanied by an increase in the number of solvent-accessible non-polar groups on the protein molecule. Such an increase is a frequent concomitant of the conformational change which characterizes denaturation. It was suggested that such proteins of increased hydrophobic character show an enhanced affinity for cell membranes and are therefore capable of acting at the membrane as signals to initiate the intracellular events, including the metabolic burst, which are necessary for a chemotactic response. The penetration of the phospholipid bilayer of the leucocyte membrane by chemoattractants may allow changes in intracytoplasmic Ca^{++} and Mg^{++} concentrations which regulate contractile events in the cell. For continued locomotion a divalent cation sequestering system in the cell would also be required. It is probable that locomotion is driven by contraction-relaxation of microfilaments. Microtubules are required to allow orientation of cells migrating in a gradient, i.e. for chemotaxis.

The binding of chemotactic factor to cell appears not to be stereospecific. We do not think that the detailed geometry of the factor is recognized by a specific receptor. It seems much more likely that recognition is mediated through hydrophobic interactions of a general type between protein and cell. It is possible that similar hydrophobic interactions are important for phagocytosis. The apparent cell-specificity of certain chemotactic factors is not understood. Further work is also required to define the role of leucocyte chemotaxis, which has been studied extensively only *in vitro*, in the defence reactions of the whole animal.

Acknowledgments

I wish to thank Dr Sally Zigmond of Yale University and Dr Ian McKay of the University of Glasgow for reading and suggesting improvements to this chapter. Part of the work described was supported by the M.R.C.

References

Allison, A. C., Davies, P. and de Petris, S. (1971). *Nature New Biology*, **232,** 153–155.

Altman, L. C. and Kirchner, H. (1972). *J. Immun.* **109,** 1149–1151.

Anfinsen, C. B. (1973). *Science*, **181,** 223–230.

Bandmann, U., Rydgren, L. and Norberg, B. (1974). *Expl. Cell Res.*, **88,** 63–73.

Becker, E. L. (1971). *In* "Biochemistry of the Acute Allergic Reactions" (K. F. Austen and E. L. Becker, eds.), pp. 243–252. Blackwell, Oxford.

Becker, E. L., Davis, A. T., Estensen, R. D. and Quie, P. G. (1972). *J. Immunol.*, **108,** 396–402.

Becker, E. L. and Showell, H. J. (1972). *Z. Immunitätsforschung*, **143,** 466–476.

Becker, E. L. and Ward, P. A. (1969). *J. exp. Med.* **129,** 569–584.

Bhisey, A. N. and Freed, J. J. (1971). *Expl. Cell Res.*, **64,** 419–429.

Borel, J. F., Keller, H. U. and Sorkin, E. (1969). *Int. Archs Allergy appl. Immun.*, **35,** 194–205.
Borel, J. F. and Sorkin, E. (1969). *Experentia*, **25,** 1333–1335.
Boyden, S. V. (1962). *J. exp. Med.* **115,** 453–466.
Breslow, E., Beychok, S., Hardman, K. D. and Gurd, F. R. N. (1965). *J. biol. Chem.* **240,** 304–309.
Brier, A. M., Snyderman, R., Mergenhagen, S. E. and Notkins, A. L. (1970). *Science*, **170,** 1104–1106.
Bryant, R. E., de Prez, R. M., Van Way, M. H. and Rogers, D. E. (1966). *J. exp. Med.* **124,** 483–499.
Buckley, I. K. (1963). *Expl. mol. Path.* **2,** 402–417.
Cagan, R. H. and Karnovsky, M. L. (1964). *Nature* (*London*), **204,** 255–257.
Carruthers, B. M. (1966). *Can. J. Physiol. Pharmacol.* **44,** 475–485.
Carruthers, B. M. (1967). *Can. J. Physiol. Pharmacol.* **45,** 269–280.
Chang, C. and Houck, J. C. (1970). *Proc. Soc. expl. Biol. Med.* **134,** 22–26.
Cohn, Z. A. (1970). *In* "Mononuclear Phagocytes" (R. van Furth, ed.), pp. 121–132. Blackwell, Oxford.
Cohn, Z. A. (1972). *In* "Inflammation, Mechanisms and Control" (I. H. Lepow and P. A. Ward, eds.), pp. 71–81. Academic Press, N.Y.
Comandon, J. (1917). *C. r. Séanc. Soc. Biol.* **80,** 314–316.
Comandon, J. (1919). *C. r. Séanc. Soc. Biol.* **82,** 1171–1174.
Estensen, R. D., Hill, H. R., Quie, P. G., Hogan, N. and Goldberg, N. D. (1973). *Nature* (*London*) **245,** 458–460.
Frédéric, J. and Robineaux, R. (1951). *J. Physiol.* (*Paris*), **43,** 732.
Goetzl, E. J. and Austen, K. F. (1974). *J. Clin. Invest.*, **53,** 591–599.
Grimes, G. J. and Barnes, F. S. (1973). *In* "Methods in Cell Biology" Vol. 6 (D. M. Prescott, ed.), pp. 325–344. Academic Press, New York.
Harris, H. (1953a). *J. Path. Bact.* **66,** 135–146.
Harris, H. (1953b). *Br. J. exp. Path.* **34,** 276–279.
Harrison, S. C. and Blout, E. (1965). *J. biol. Chem.* **240,** 299–303.
Hill, J. H. and Ward, P. A. (1969). *J. exp. Med.* **130,** 505–518.
Holland, P., Holland, N. H. and Cohn, Z. A. (1972). *J. exp. Med.* **135,** 458–475.
Houck, J. C. and Chang, C. (1971). *Proc. Soc. expl. Biol. Med.* **138,** 69–75.
Iyer, G. Y. N., Islam, M. F. and Quastel, J. H. (1961). *Nature* (*London*), **192,** 535–541.
Javaherian, K. and Beychok, S. (1968). *J. molec. Biol.* **37,** 1–11.
Jensen, J. A., Snyderman, R. and Mergenhagen, S. E. (1969). *In* "Cellular and Humoral Mechanisms in Anaphylaxis and Allergy" (H. Z. Movat, ed.), p. 265. Karger, Basel.
Jensen, J. A. and Esquenazi, V. (1975). *Nature* (*London*) **256,** 213–215.
Kaley, G. and Weiner, R. (1971). *Nature New Biology*, **234,** 114–115.
Kaplan, A. P., Kay, A. B. and Austen, K. F. (1972). *J. exp. Med.* **135,** 81–97.
Karnovsky, M. L., Simmons, S., Glass, E. A., Shafer, A. W. and D'Arcy Hart, P. (1970). *In* "Mononuclear Phagocytes" (R. van Furth, ed.), pp. 103–120. Blackwell, Oxford.
Kauzmann, W. (1959). *Adv. Protein. Chem.* **14,** 1–63.
Kay, A. B. (1970). *Clin. expl. Immun.* **7,** 723–737.

KAY, A. B. and AUSTEN, K. F. (1972). *Clin. Expl. Immun.* **11,** 557–563.
KAY, A. B., PEPPER, D. S. and EWART, M. R. (1973). *Nature New Biology,* **243,** 56–57.
KAY, A. B., SHIN, H. S. and AUSTEN, K. F. (1973). *Immunology,* **24,** 969–976.
KAY, A. B., STECHSCHULTE, D. J. and AUSTEN, K. F. (1971). *J. exp. Med.* **133,** 602–619.
KELLER, H. U., BOREL, J. F., WILKINSON, P. C., HESS, M. and COTTIER, H. (1972). *J. immun. Methods,* **1,** 165–168.
KELLER, H. U. and SORKIN, E. (1966). *Immunology* **10,** 409–416.
KELLER, H. U. and SORKIN, E. (1967a). *Int. Archs Allergy appl. Immun.* **31,** 505–517.
KELLER, H. U. and SORKIN, E. (1967b). *Int. Archs Allergy appl. Immun.* **31,** 575–586.
KNIGHT, B. (1974). Appendix to paper by S. H. Zigmond (*op. cit.,* 1974a).
LANDSTEINER, K. (1936). "The specificity of serological reactions". Charles C. Thomas, Springfield, Illinois (revised edition, 1962), Dover, New York.
LAY, W. H. and NUSSENZWEIG, V. (1968). *J. exp. Med.* **128,** 991–1009.
LEAHY, D. R., MCLEAN, E. R. and BONNER, J. T. (1970). *Blood,* **36,** 52–54.
MARCHESI, V. T. and PALADE, G. E. (1967). *J. Cell. Biol.* **35,** 385.
MCCUTCHEON, M. (1946). *Physiol. Rev.* **26,** 319–336.
MCKENZIE, H. A. (ed.), (1971). "Milk Proteins: Chemistry and Molecular Biology", Vol. II. Academic Press, New York.
MUDD, E. B. H. and MUDD, S. (1933). *J. gen. Physiol.* **16,** 625–636.
MUDD, S. and MUDD, E. B. H. (1924). *J. exp. Med.* **40,** 647–660.
NYDEGGER, U. E., ANNER, R. M., GEREBTZOFF, A., LAMBERT, P. H. and MIESCHER, P. A. (1973). *Eur. J. Immun.* **3,** 465–470.
OSS, C. J. VAN and GILLMAN, C. F. (1972). *J. reticuloendothelial Soc.* **12,** 283–292.
PARISH, W. E. (1972). *Clinical Allergy,* **2,** 381–390.
RABINOVITCH, M. (1970). *In* "Mononuclear Phagocytes" (R. van Furth, ed.), pp. 299–315. Blackwell, Oxford.
RAMSEY, W. S. (1972a). *Expl. Cell Res.* **70,** 129–139.
RAMSEY, W. S. (1972b). *Expl Cell Res.* **72,** 489–501.
ROBINEAUX, R. (1954). *Revue Hémat.* **9,** 364–402.
ROBINEAUX, R. (1964). *In* "Primitive motile systems in cell biology" (R. D. Allen and N. Kamiya, eds.), pp. 351–364. Academic Press, New York.
ROBINEAUX, R. and FRÉDÉRIC, J. (1955). *C. r. Séanc. Soc. Biol.* **149,** 486–489.
ROSSI FANELLI, A., ANTONINI, E. and CAPUTO, A. (1958). *Biochim. biophys. Acta* **30,** 608–615.
RUSSELL, R. J., MCINROY, R. J., WILKINSON, P. C. and WHITE, R. G. (1975). Behring Institute Mitteilungen, **57,** 103–109.
SBARRA, A. J. and KARNOVSKY, M. L. (1959). *J. biol. Chem.* **234,** 1355–1362.
SINGER, S. J. and NICOLSON, G. L. (1972). *Science,* **175,** 720–731.
SNYDERMAN, R. (1973). *Nouv. Revue fr. Hémat,* **13,** 888 (abstract).
SNYDERMAN, R., GEWURZ, H. and MERGENHAGEN, S. E. (1968). *J. exp. Med.* **128,** 259–275.
SNYDERMAN, R., SHIN, S. H. and HAUSMAN, M. S. (1971). *Proc. Soc. exp. Biol. Med.* **138,** 387–390.

SNYDERMAN, R., SHIN, S. H., PHILLIPS, J. K., GEWURZ, H. and MERGENHAGEN, S. E. (1969). *J. Immun.* **103,** 413–422.

SORKIN, E., STECHER, V. J. and BOREL, J. F. (1970). *Series haematologica,* **3**(1), 131–162.

STECHER, V. J. and SORKIN, E. (1972). *Int. Archs Allergy appl. Immun.* **43,** 879–886.

STECHER, V. J., SORKIN, E. and RYAN, G. B. (1971). *Nature New Biology,* **233,** 95–96.

STOSSELL, T. and POLLARD, T. D. (1973). *J. biol. Chem.,* **248,** 8288–8294.

SYMON, D. N. K., MCKAY, I. C. and WILKINSON, P. C. (1972). *Immunology,* **22,** 267–276.

TATSUMI, N., SHIBATA, N., OKAMURA, Y., TAKEUCHI, K. and SENDA, N. (1973). *Biochem. biophys. Acta,* **305,** 433–444.

THRASHER, S. G., YOSHIDA, T., VAN OSS, C. J., COHEN, S. and ROSE, N. R. (1973). *J. Immun.* **110,** 321–326.

TSE, R. L., PHELPS, P. and URBAN, D. (1972). *J. lab. clin. Med.* **80,** 264–274.

WARD, P. A. and BECKER, E. L. (1968). *J. exp. Med.* **127,** 693–709.

WARD, P. A. and BECKER, E. L. (1970a). *J. Immun.* **105,** 1057–1067.

WARD, P. A. and BECKER, E. L. (1970b). *Life Sciences, part* 2, **9,** 355–360.

WARD, P. A., COCHRANE, C. G. and MÜLLER-EBERHARD, H. J. (1966). *Immunology,* **11,** 141–153.

WARD, P. A., COHEN, S. and FLANAGHAN, T. D. (1972). *J. exp. Med.* **135,** 1095–1103.

WARD, P. A. and HILL, J. H. (1970). *J. Immun.* **104,** 535–543.

WARD, P. A. and NEWMAN, L. J. (1969). *J. Immun.* **102,** 93–99.

WARD, P. A., REMOLD, H. G. and DAVID, J. R. (1969). *Science,* **163,** 1079–1081.

WAUGH, D. F., CREAMER, L. K., SLATTERY, D. W. and DRESDNER, G. W. (1970). *Biochemistry,* **9,** 786–795.

WILKINSON, P. C. (1972). *Experientia,* **28,** 1051–1052.

WILKINSON, P. C. (1973). *Nature (London),* **244,** 512–513.

WILKINSON, P. C. (1974a). "Chemotaxis and Inflammation". Churchill-Livingstone, Edinburgh.

WILKINSON, P. C. (1974b). *Nature (London),* **251,** 58–60.

WILKINSON, P. C. (1975). *Expl. Cell Res.,* in press.

WILKINSON, P. C., BOREL, J. F., STECHER-LEVIN, V. J. and SORKIN, E. (1969). *Nature (London),* **222,** 244–247.

WILKINSON, P. C. and MCKAY, I. C. (1971). *Int. Archs Allergy appl. Immun.* **41,** 237–247.

WILKINSON, P. C. and MCKAY, I. C. (1972). *Eur. J. Immun.* **2,** 570–577.

WILKINSON, P. C. and MCKAY, I. C. (1974). *In* "Chemotaxis: Its Biology and Biochemistry" (E. Sorkin, ed.), pp. 421–441. Karger, Basel.

WILKINSON, P. C., O'NEILL, G. J., MCINROY, R. J., CATER, J. C. and ROBERTS, J. A. (1973a). *In* "Immunopotentiation" (G. E. Wolstenholme and J. Knight, eds.), p. 121. Associated Scientific Publishers, Amsterdam.

WILKINSON, P. C., O'NEILL, G. J. and WAPSHAW, K. (1973b). *Immunology,* **24,** 997–1006.

WISSLER, J. H. and SORKIN, E. (1973). *Nouv. Revue. fr. Hémat,* **13,** 896 (abstract).

WISSLER, J. H., STECHER, V. J. and SORKIN, E. (1972). *Int. Archs. Allergy appl. Immun.* **42,** 722–747.
WOODIN, A. M. (1971). *In* "The Reticuloendothelial System and Immune Phenomena" (N. A. di Luzio and K. Flemming, eds.), p. 71. Plenum Press, New York.
WOODIN, A. M. and HARRIS, A. (1973). *Expl. Cell Res.* **77,** 41–46.
WOODIN, A. M. and WIENEKE, A. A. (1969). *Brit. J. exp. Path.* **50,** 295–308.
WOODIN, A. M. and WIENEKE, A. A. (1970). *Nature (London)*, **227,** 460–463.
YOSHINAGA, M., YAMAMOTO, S., KIYOTA, S. and HAYASHI, H. (1972). *Immunology*, **22,** 393–400.
ZIGMOND, S. H. (1974a). *In* "Chemotaxis: Its Biology and Biochemistry" (E. Sorkin, ed.), in press. Karger, Basel.
ZIGMOND, S. H. (1974b). *Nature (London)*, **249,** 450–452.
ZIGMOND, S. H. and HIRSCH, J. G. (1972). *Expl. Cell Res.* **73,** 383–393.
ZIGMOND, S. H. and HIRSCH, J. G. (1973). *J. exp. Med.* **137,** 387–410.

Index

D

T